Holt Chemistry
Visualizing Matter

LABORATORY EXPERIMENTS
TEACHER'S EDITION

HOLT, RINEHART AND WINSTON
Harcourt Brace & Company

Austin • New York • Orlando • Atlanta • San Francisco • Boston • Dallas • Toronto • London

Lab Authors

Dave Jaeger
Chemistry Teacher
Will C. Wood High School
Vacaville, CA

Suzanne Weisker
Science Teacher and Department Chair
Will C. Wood High School
Vacaville, CA

Safety and Disposal Reviewer

Jay A. Young, Ph.D.
Chemical Safety Consultant
Silver Spring, MD

Teacher Reviewers

Cheryl Epperson
Chemistry Teacher
Flour Bluff High School
Corpus Christi, TX

Marilyn Lawson
Chemistry Teacher/Science Department Chair
Gregory Portland High School
Portland, TX

Cover: (coal) Telegraph Colour Library/FPG International Corp., (diamond structure, orbital) Foca, (diamonds) © 1991 Martin Rogers/FPG International Corp., (full cover wrap) © Goavec Photography/The Image Bank
Illustrations: Progressive Information Technologies

Printed in the United States of America

ISBN 0-03-095285-9

23456 129 98 97 96 95

Table of Contents

For best results and the greatest safety, do not perform any *Investigation* without performing its prerequisite *Exploration*, if it has one. The prerequisite is identified at the beginning of the Teacher's Notes for each *Investigation*.

Laboratory Program Overview

Each lab exercise takes place in the context of a real-world scenario. As a result, students will be able to see how the techniques and concepts they are learning in the lab and classroom can relate to problems in the real world.

These labs take concepts into the real world

An integral part of the context of each lab's scenario is the role of the student as an employee of a scientific consulting firm. All students will experience what it is like to be an employee, to follow directions, and to apply creative thinking to solve problems with available resources.

Two types of labs

The lab exercises in the textbook and the lab manual are divided into two types. *Explorations* have complete procedures provided for students, along with thorough questions designed to help students make sense of what they observed in the lab. *Investigations* are complete scenarios that allow for student-designed procedures and open-ended lab work. Most of these labs are in pairs, with the *Exploration* serving as a precursor to an accompanying *Investigation*.

> **For best results and the greatest safety, do not perform any *Investigation* without performing its prerequisite *Exploration*, if it has one. The prerequisite is identified at the beginning of the Teacher's Notes for each *Investigation*.**

Explorations: Technique Builders with Procedures

Nearly all *Investigations* are preceded by an *Exploration*, which serves as a *Technique Builder* by giving students practice with lab techniques and concepts. Each *Exploration* includes a detailed, step-by-step procedure, like many traditional labs. Unlike labs from other lab programs, the *Explorations* are placed in the context of a scenario. A situation is described, and students are given a

> **A more thorough explanation of the organization of each type of lab and its potential uses can be found in the annotated teacher's edition of the textbook. If you decide to try these labs, you should start by examining the 42 lab exercises found in the textbook. This lab manual and its 32 additional lab exercises should be used as a supplement because many of the lab exercises in this laboratory manual require the use of more complicated equipment or more difficult procedures than do those in the textbook.**

problem to solve. The procedure and the items that follow it lead students through the techniques and thought processes that are necessary to solve the problem.

Investigations: Open-ended Problem Solving

The *Investigation* that follows an *Exploration* gives students a chance to use their skills in *Problem Solving* as they examine another aspect of the technique, reaction, or concept that they encountered in the *Exploration*. Instead of instructions, students are given a business letter from a "client" that outlines a problem and a memo from their supervisor that offers further suggestions. Based on the problem, they must create a plan for their procedure, select equipment from a list of available materials, and attempt to stay within a budget. Student plans are reviewed by you, the teacher, for practicality and safety before the students enter the lab. On completion of the lab, instead of writing a traditional lab report, students write a business letter to the "client" explaining their results and submitting a bill for their services.

A program that maximizes flexibility

Of course, you, the teacher, are the only person who knows what is best for your situation. The lab program was designed to be flexible so that you can customize it to fit your situation and your students' needs. You may not be certain whether the open-ended, student-designed nature of the

Investigations is right for you and your students. If you decide not to use the *Investigations* with your students, you still have 37 *Explorations* to use, 19 in the textbook and another 18 in the lab manual.

For many teachers, there will be a variety of logistical issues to work out before they try student-designed labs. Because each *Exploration-Investigation* pair stands alone, you can choose any pair as the one you will try with your students. The authors strongly recommend that if you haven't tried this type of lab experience before, you should *start slowly, with only a few open-ended labs spread throughout the year.* Be sure to allow extra time for the students to design, refine, and perfect their procedures. After you have grown comfortable with the approach, add more in successive years.

What shortcuts can I implement to simplify things?

Even within individual *Investigations,* there are numerous adaptations that can be made to streamline the process. One possibility is to delete the budgeting and invoice tasks. Another possibility is to tell the students exactly what equipment they need instead of having them decide. A concise list is provided in the Teacher's Notes for each *Investigation.* You could even tell students part of the proposed procedure provided in the Teacher's Notes, instead of requiring them to decide on a plan themselves.

Tips for carrying out Explorations

Most of the techniques you use with traditional labs will work when you assign an *Exploration.* Be sure to assign the following tasks for students to complete before entering the lab.

- Read the *Situation, Background, Problem, Safety, Preparation,* and *Technique* sections.
- Re-read the *Safety* section.
- Create fully prepared data tables according to the instructions given in the *Preparation* section.

You may even want to give students a brief quiz on the safety precautions or the procedure. Only those with a perfect score should be allowed to proceed.

Reports

After the *Exploration,* students should prepare a **lab report** with at least the following components.

- **title**
- **summary paragraph** describing the purpose and procedure
- **data tables and observations** that are organized and comprehensive
- **answers** to the *Analysis and Interpretation, Conclusions,* and any *Extensions* items you assign

Be sure to instruct students to give fully worked-out answers. It may help your grading if you also tell them to circle, box, or highlight the final result of their calculations. If you will require other components for the lab reports, be certain to explain them to students.

A scoring rubric for *Exploration* lab reports is given on page T9.

Tips for carrying out Investigations

The *Investigations* require an approach that is different from most traditional labs. Students at this stage tend to panic when left on their own to develop a lab, particularly a procedure. Be sure to do the following to help them feel comfortable.

- **Completely discuss the *Exploration* prerequisite.**
- **Suggest that students use equipment and amounts similar to those in the *Exploration*** if they are uncertain about how to proceed.
- **Remind students to concentrate on what they need to know and how they will measure it** so that they remain focused.
- **Remind students that it is more important to understand** what is going on in the lab than to perform it with excellent technique but no understanding.
- **Provide leading questions for students to consider** as they make their plans. Suggestions are often given in the *Pre-Lab Discussion* section of the Teacher's Notes for the *Investigations.*
- **Hold a question-and-answer session** before students begin the lab.
- **Hint (don't tell) that some of the equipment listed may be unnecessary.**
- **Check the memo in the *Investigation*** for details on the requirements for the preliminary report.

The Teacher's Notes for each *Investigation* contain tips that will help you evaluate the pre-lab requirements. A concise list details all of the equipment and materials that each lab group should request, along with the projected total cost. A proposed procedure is also included so that you can compare it to students' suggested plans and promptly notice any discrepancies. Be sure to tell students when there are flaws. You may require them to propose solutions themselves, instead of telling them what to do.

Reports

The **preliminary report** is an essential part of these labs. If done properly, the actual procedure of the lab can usually be completed in one period. If students are disorganized, they may need to purchase more lab-space time (if you permit it) either during class or after school.

A scoring rubric for *Investigation* preliminary reports is given on page T12.

On completion of the lab, students should report their results in the form of a **business letter.** Specific requirements are indicated in the memo on the second page of the student's *Investigation.* Be sure to emphasize to students that the goal is to communicate their results clearly and concisely. Students should also submit an invoice.

A scoring rubric for the *Investigation* business letter is given on page T15.

Microcomputer-based laboratory

Some *Explorations* and *Investigations* include procedures for using equipment such as pH meters or LEAP System microcomputer-based laboratory probes from Quantum Technology. Each time this option is presented, an alternative procedure that does not require such equipment is also presented. Thus, the lab activities remain useful and practical whether the technology is available or not.

To order the LEAP System and probes, contact Quantum Technology. Quantum has set up a variety of packages for users of *Holt Chemistry: Visualizing Matter.*

Quantum Technology
30153 Arena Drive
Evergreen, CO 80439

Sales and Technical Support: (303) 674-9651
Fax: (303) 674-6763

Lab	Title	LEAP System option
Inv 6-1 Exp. 6-2 Inv. 6-2	Covalent and Ionic Compounds Viscosity of Liquids	conductivity probe thermistor probe
Exp. 7-1	Chemical Reactions and Solid Fuel	thermistor probe
Exp. 9-1 Inv. 9-1	Specific Heat Capacity	thermistor probe
Exp. 9-2 Inv. 9-2	Constructing Heating/Cooling Curves	thermistor probe
Exp. 9-3	Heat of Fusion	thermistor probe
Exp. 9-4	Heat of Solution	thermistor probe
Inv. 10-1	Gas Temperature-Volume Relationship	thermistor probe
Exp. 11-2	Testing for Dissolved Oxygen	dissolved oxygen probe
Exp. 11-3	Solubility of Ammonia	thermistor probe
Exp. 12-2 Inv. 12-2	Freezing-Point Depression	thermistor probe
Inv. 13-1 Inv. 13-2	Measuring pH	pH probe
Inv. 13-3	Acid-Base Titration	pH probe
Exp. 15-2 Inv. 15-2	Voltaic Cells	voltage probe

Evaluating Lab Work—Scoring Rubrics

Traditional means of scoring labs on a percentage scale may not work well with this type of lab program. Of course, you are the only one who can judge what is best for your students. The following scoring rubrics, which are based on a scale of 0–6, may be a useful starting point. Read the description of each level, and decide which one most accurately fits each report you grade. Although the first few that you grade may be time-consuming, as you grow accustomed to the rubric, it will soon become a very quick process.

A useful strategy is to keep a file of past papers that seem to exemplify each level to use as benchmarks. You may want to make this file available to students to give them examples of good work. Examples of Investigation business letter reports and invoices from the author's classes are found on pages T18–T26.

Exploration Lab Report

Technique Builder

Experienced Level (6 points)
- Excellent technique was used throughout the lab procedure.
- Data and observations were recorded accurately, descriptively, and completely, with no serious errors.
- *Analysis and Interpretation* items were performed clearly, concisely, and accurately, with correct units and properly worked-out calculations.
- Graphs, if necessary, are drawn accurately and neatly.
- Students express their recognition of the connections between their observations and the related chemistry concepts in an exemplary manner.
- Good reasoning and logic are evident throughout the report.
- Answers to *Conclusions* items are written correctly and accurately.
- Any *Extensions* items that were assigned are completed with creativity and imagination.

Competent Level (5 points)
- No errors in technique were observed during the lab procedure.
- Data and observations were recorded accurately, descriptively, and completely, with only minor errors.
- *Analysis and Interpretation* items were performed accurately, with correct units and properly worked-out calculations, but may have been slightly disorganized.
- Graphs, if necessary, are drawn accurately and neatly.
- Students effectively express their recognition of the connections between their observations and the related chemistry concepts.
- Good reasoning and logic are evident throughout the report.
- Answers to *Conclusions* items are written correctly and accurately, but there may be minor misunderstandings.
- Any *Extensions* items that were assigned are completed adequately.

Intermediate Level (4 points)

- Only minor errors in technique were observed during the lab procedure.
- Data and observations were recorded accurately with only minor errors or omissions.
- *Analysis and Interpretation* items were performed accurately, but some minor errors were made either in calculations or in applying correct units.
- Graphs, if necessary, are drawn accurately and neatly.
- Students satisfactorily express their recognition of the connections between their observations and the related chemistry concepts.
- Reasoning is occasionally weak in the report, but only in a few places.
- Answers to *Conclusions* items are correct, but there are some misunderstandings or minor errors.
- Students made an effort to adequately address any *Extensions* items that were assigned, but their effort may not have been entirely successful.

Transitional Level (3 points)

- Only a few errors in technique were observed during the lab procedure, but they may have been significant.
- Data and observations were recorded accurately, with only minor errors or omissions.
- *Analysis and Interpretation* items were performed accurately, but some minor errors were made both in calculations and in applying correct units.
- Graphs, if necessary, are drawn accurately and neatly.
- Students recognize connections between their observations and the related chemistry concepts, but only weakly express their understanding.
- Reasoning is generally weak throughout much of the report.
- Some answers to *Conclusions* items are not correct because of misunderstandings or minor errors.
- Students made an effort to address any *Extensions* items that were assigned, but their effort was not substantial enough.

Exploration Lab Report

Technique Builder

Beginning Level (2 points)

- Several serious errors in technique were observed during the lab procedure.
- Most data and observations were recorded accurately, but with several significant errors or omissions.
- *Analysis and Interpretation* items were performed inaccurately, but correct units were used most of the time.
- Graphs, if necessary, are drawn adequately.
- Students may or may not recognize the connections between their observations and the related chemistry concepts, but they don't express this understanding in the report.
- Errors in logic are made in the report.
- Answers to *Conclusions* items are incorrect or poorly written.
- Students did not make a meaningful effort to address any *Extensions* items that were assigned.

Inexperienced Level (1 point)

- Many serious errors in technique were observed during the lab procedure.
- Data and observations are inaccurate or incomplete.
- *Analysis and Interpretation* items were performed inaccurately, with no units or incorrect ones.
- Graphs, if necessary, are drawn incorrectly.
- Students clearly do not recognize the connections between their observations and the related chemistry concepts.
- Errors in logic are made throughout the report.
- Answers to *Conclusions* items are so incorrect that it is obvious the students did not understand the lab.
- Students made no effort at all to address any *Extensions* items that were assigned.

Unacceptable Level (0 points)

- All work is unacceptable.
- No responses are relevant to lab.
- Major components of lab report are missing.

**Problem
Solving**

Experienced Level (6 points)

- Plan shows careful and thorough planning with good reasoning and logic.

- Plan is complete, appropriate, and safe.

- Plan is as efficient as the procedure given in the Teacher's Notes for the *Investigation*.

- Plan is expressed clearly and concisely.

- Proposed data tables are made properly and clearly indicate all measurements that must be made to solve the problem.

- Proposed list of equipment includes all equipment necessary to carry out the procedure described in the plan.

- When compared to the list given in the Teacher's Notes for the *Investigation*, the student's list of equipment includes no unnecessary pieces of equipment.

- Proposed budget is within $10 000 of the materials' cost listed in the Teacher's Notes for the *Investigation*.

- Any other requirements listed in the memorandum for the *Investigation* are addressed completely, clearly, and concisely.

Competent Level (5 points)

- Plan shows careful and thorough planning, although the logic behind it may not be clearly expressed.

- Plan is appropriate, safe, and mostly complete, with only minor omissions.

- Plan is almost as efficient as the procedure given in the Teacher's Notes for the *Investigation*.

- Some parts of the plan could be expressed more clearly or more concisely.

- Proposed data tables indicate all measurements that must be made to solve the problem, but minor errors may have been made in preparing the data tables.

- Proposed list of equipment includes all equipment necessary for the procedure described in the plan.

- When compared to the list given in the Teacher's Notes for the *Investigation*, the student's list of equipment may include one or two unnecessary pieces of equipment.

- Proposed budget is within $20 000 of the materials' cost listed in the Teacher's Notes for the *Investigation*.

- Any other requirements listed in the memorandum for the *Investigation* are completely addressed, but they could be more clear or more concise.

Evaluating Lab Work—Scoring Rubrics, continued

Problem Solving

Intermediate Level (4 points)

- Plan shows some logic, but the reasoning could have been more careful, more thorough, or more clearly expressed.
- Plan is appropriate and safe, but there are a few omissions.
- Plan will work, but is not as efficient as that given in the Teacher's Notes for the *Investigation*.
- Plan could be written more clearly or more concisely.
- Proposed data tables indicate all measurements that must be made to solve the problem, but multiple trials are not included.
- Proposed list of equipment includes almost all equipment necessary to carry out the plan described in the procedure, but there may be minor omissions.
- When compared to the list given in the Teacher's Notes for the *Investigation*, the student's list of equipment may include a few unnecessary pieces of equipment.
- Proposed budget is within $30 000 of the materials' cost listed in the Teacher's Notes for the *Investigation*.
- Any other requirements listed in the memorandum for the *Investigation* are addressed, but not completely.

Transitional Level (3 points)

- Plan shows some logic, but not enough to completely solve the problem.
- Plan is safe, but it includes inappropriate procedures or omits necessary steps.
- Plan will probably not work and does not match the procedure given in the Teacher's Notes for the *Investigation*.
- Plan is poorly written.
- Proposed data tables include most of the necessary information, but errors have been made in preparing them.
- Proposed list of equipment omits a few pieces of equipment necessary for the procedure described in the plan.
- When compared to the list given in the Teacher's Notes for the *Investigation*, the student's list of equipment may include several unnecessary pieces of equipment.
- Proposed budget is within $50 000 of the materials' cost listed in the Teacher's Notes for the *Investigation*.
- One of the other requirements listed in the memorandum for the *Investigation* is not addressed completely.

INVESTIGATION

Problem Solving

Beginning Level (2 points)

- Plan shows only a small amount of logic or understanding of what is necessary to solve the problem.
- Plan may not be completely safe.
- Plan will definitely not work and does not match the procedure given in the Teacher's Notes for the *Investigation*.
- Plan is poorly written.
- Proposed data tables include some necessary information, but there are some serious omissions.
- Proposed list of equipment has so many omissions that it would be difficult to carry out the procedure described in the plan.
- When compared to the list given in the Teacher's Notes for the *Investigation*, the student's list of equipment includes many unnecessary pieces of equipment.
- Proposed budget is within $70 000 of the materials' cost listed in the Teacher's Notes for the *Investigation*.
- Few of the other requirements listed in the memorandum for the *Investigation* are addressed completely.

Inexperienced Level (1 point)

- Plan does not clearly show logic or understanding of the problem.
- Plan is unsafe.
- Plan will definitely not work and does not match the procedure given in the Teacher's Notes for the *Investigation*.
- Plan is poorly written.
- Proposed data tables include only a few of the necessary pieces of data.
- Proposed list of equipment has so many omissions that it would be impossible to carry out the procedure described in the plan.
- Not enough critical thinking has gone into the choice of equipment; most of the items on the materials list given in the *Investigation* are included, whether necessary or not.
- Proposed budget is within $100 000 of the materials' cost listed in the Teacher's Notes for the *Investigation*.
- None of the other requirements listed in the memorandum for the *Investigation* are addressed completely.

Unacceptable Level (0 points)

- All work unacceptable.
- Major components of the preliminary report are missing.
- Plan is completely illogical or unsafe, or large portions are omitted.
- Plan is poorly written.
- Proposed data tables are entirely unsatisfactory.
- No thought has been given to the choice of equipment; all of the items on the materials list given in the *Investigation* are included.
- Proposed budget is more than $100 000 over or under the materials' cost listed in the Teacher's Notes for the *Investigation*.

Investigation Business Letter including Invoice

Problem Solving

Experienced Level (6 points)

- Excellent technique was used throughout the lab procedure.
- The letter clearly expresses good reasoning, logic, and understanding of what the lab was about.
- The connections between the data and the related chemistry concepts that lead to the conclusions are clearly expressed.
- Students recognize the connection between the initial problem and the outcome of the lab work.
- Summary is clear, concise, and well organized, with few grammatical or stylistic errors.
- Data and analysis are accurate and well organized, with correct units and no errors in calculations.
- Percent error for quantitative answers is less than 15%.
- Graphs, if appropriate, are drawn and labeled accurately and neatly.
- Any other requirements listed in the memorandum for the *Investigation* are addressed completely, clearly, and concisely.

Competent Level (5 points)

- No errors in technique were observed during the lab procedure.
- The letter expresses basically sound reasoning, logic, and understanding of what the lab was about, although there may be some minor errors.
- The connections between the data and the related chemistry concepts that lead to the conclusions are expressed, but not clearly.
- Students recognize the connection between the initial problem and the outcome of the lab work.
- Summary is essentially clear, concise, and well organized, with no serious grammatical or stylistic errors.
- Data and analysis are accurate, with correct units and no errors in calculations but they may not be well-organized.
- Percent error for quantitative answers is less than 25%.
- Graphs, if appropriate, are drawn and labeled accurately and neatly.
- Any other requirements listed in the memorandum for the *Investigation* are completely addressed, but they could be more clear or more concise.

INVESTIGATION

Problem Solving

Intermediate Level (4 points)

- Only minor errors in technique were observed during the lab procedure.
- The letter expresses some logic and understanding of what the lab was about, although there are several minor errors.
- The connections between the data and the related chemistry concepts that lead to the conclusions are not clearly expressed, but students seem to understand them.
- Students recognize the connection between the initial problem and the outcome of the lab work.
- Summary is essentially correct, but it is poorly organized or contains serious grammatical and stylistic errors.
- Data and analysis are mostly accurate but are poorly organized or have units that are incorrect or some calculations that are not shown in full detail.
- Percent error for quantitative answers is less than 35%.
- Graphs, if appropriate, are drawn and labeled accurately but are not neat.
- Any other requirements listed in the memorandum for the *Investigation* are addressed, but not completely.

Transitional Level (3 points)

- Only a few errors in technique were observed during the lab procedure, but they may have been significant.
- The letter expresses some logic but little understanding of what the lab was about.
- The connections between the data and the related chemistry concepts that lead to the conclusions are not clearly expressed, and students do not seem to understand them completely.
- Students do not recognize the connection between the initial problem and the outcome of the lab work.
- Summary contains a few minor errors of fact along with grammatical and stylistic errors.
- Data and analysis are poorly organized or inaccurate because of minor errors, or calculations are not shown in full detail.
- Percent error for quantitative answers is less than 50%.
- Graphs, if appropriate, are drawn accurately but are either mislabeled or not neat.
- One of the other requirements listed in the memorandum for the *Investigation* is not addressed completely.

Investigation Business Letter including Invoice

Problem Solving

Beginning Level (2 points)

- Several serious errors in technique were observed during the lab procedure.
- The letter expresses little logic, and there seems to be little understanding of what the lab was about.
- The connections between the data and the related chemistry concepts that lead to the conclusions are clearly not understood.
- Students do not recognize the connection between the initial problem and the outcome of the lab work.
- Summary contains several errors of fact or is poorly written.
- Data and analysis contain a few significant errors or are poorly organized, or calculations are not shown at all.
- Percent error for quantitative answers is less than 65%.
- Graphs, if appropriate, are inaccurate or incomplete.
- Few of the other requirements listed in the memorandum for the *Investigation* are addressed completely.

Inexperienced Level (1 point)

- Many serious errors in technique were observed during the lab procedure.
- The letter expresses little logic, and there seems to be no understanding of what the lab was about.
- There is no attempt to connect the data and the related chemistry concepts to lead to a conclusion.
- Students do not recognize the connection between the initial problem and the outcome of the lab work.
- Summary contains many errors of fact or is poorly written.
- Data and analysis contain many significant errors or are poorly organized, or calculations are not shown at all.
- Percent error for quantitative answers is less than 80%.
- Graphs, if appropriate, are missing.
- None of the other requirements listed in the memorandum for the *Investigation* are addressed completely.

Unacceptable Level (0 points)

- Many serious errors in technique were observed during the lab procedure.
- All work is unacceptable.
- Few responses are relevant to the lab.
- Major components of the letter are missing.
- The report expresses little logic, and there seems to be no understanding of what the lab was about.
- There is no attempt to connect the data and the related chemistry concepts to lead to a conclusion.
- Students do not recognize the connection between the initial problem and the outcome of the lab work.
- Summary contains many errors of fact or is poorly written.
- Data and analysis are incomplete and contain many significant errors.
- Percent error for quantitative data is more than 80%.

Evaluating Lab Work—Sample Reports

The following papers are slightly revised versions of reports written by students in one author's classrooms. They are placed here to provide you with benchmarks that demonstrate the use of the six-point scoring rubric outlined earlier.

The grade that these papers would receive in your class may vary from the grade given by the author. What is most important is that your use of the rubric is consistent within your classes. In this way, you can be certain that you are fair in comparing students' work. At the same time, the students learn what expectations you will have for reports in your class.

These papers, which were written before the completion of this lab manual, do not always perfectly match the structure suggested by the book. For example, rather than CheMystery Labs of Springfield, Virginia, the author encourages students to set up details and particulars about their own chemical analysis firm, including its name, address, and other details. However, these reports are still a useful supplement to the scoring rubrics, because they demonstrate the flexibility of its use, as well as how readily it can be customized to fit the specific needs of your class.

You may have your own plans and ideas about how to grade such reports. The scoring rubrics and these sample papers are not meant to dictate how grading should be done, but rather to provide some helpful suggestions as you develop a system that works for your situation and with your students. Your final system may be very similar to or very different from the approach used by the authors.

Some teachers will want to make copies of these sample reports for students to examine and use as models. Rather than giving every student a copy, the authors suggest that you keep a file of these papers for students to consult. Otherwise, students tend to slavishly follow the form and language of the sample reports, instead of thinking carefully and being creative and original. Throughout the years that you use this lab program, copies of reports from students in your class that exemplify each scoring level of the rubric can be added to this file.

Superior California Analytical Laboratories
7413 Mountain View Parkway
Vacaville, CA
95687

October 17, 1994

Maxwell Stone
Director of Anti-Terrorist Operations—North American Theater
CIA Headquarters
Langley, Virginia

Dear Mr. Stone:

The cause of the crash of Crest Air flight 998 was an explosive, probably planted by the terrorist organization Spleen Activist Group. A piece of metal recovered from the bomb has been identified as aluminum, which is the metal of preference for the Spleen Activist Group's devices.

Sample #903 was identified using the intensive properties of density and specific heat capacity. The average density of the sample was calculated to be 2.80 g/mL, which is almost identical to the density of aluminum (2.69 g/mL). The average specific heat capacity of sample #903 was calculated at 0.816 J/g•C°, while the known specific heat capacity of aluminum is 0.897 J/g•C°. These results demonstrate that sample #903 is aluminum.

The volume of sample 903 was ascertained in two ways. Sample #903 was a cylinder, so volume was calculated using the following formula (Trial 4).

$$Volume = Radius^2 \times \pi \times Height$$

For the first three trials, the volume was measured using water displacement. Sample #903 was added to 15 mL of water in a 25 mL graduated cylinder, causing the water level in the cylinder to rise. The difference between the original water level and the new water level was then computed, determining the volume of sample #903. The metal sample was then dried and its mass measured. The density of the sample was calculated using the following equation.

$$Density = \frac{Mass}{Volume}$$

Along with density, three trials measuring specific heat capacity were conducted and used to confirm the identity of sample #903. The specific heat was ascertained using a Styrofoam calorimeter and the following formula:

$$Spec.\ heat\ capacity\ metal = \frac{Mass\ H_2O \times \Delta t\ H_2O \times Spec.\ heat\ capacity\ H_2O}{Mass\ metal \times \Delta t\ metal}$$

Author's Example of Experienced Level (6 points)

The following chart contains the data for specific heat capacity (J/g•°C) and density (g/mL), along with average deviation and average percent error of the trials.

Tables of Values for Density and Specific Heat

	Trial 1	Trial 2	Trial 3	Trial 4	Trial Avg.	Avg. Dev.	Avg. % Error
Density (g/mL)	2.83	2.83	2.83	2.71	2.80	0.045	4.1
Specific Heat Capacity (J/g•°C)	0.878	0.735	0.834		0.816	0.054	9.0

At right is a graph comparing the average density of sample #903 to the known densities of aluminum, iron, lead, tin, and zinc. The graph explicitly demonstrates the similarities between the sample and aluminum.

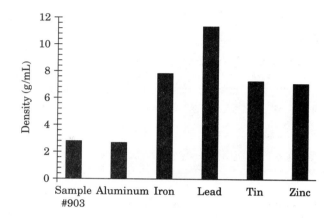

The graph at right displays the average specific heat of sample #903 compared to aluminum, iron, lead, tin, and zinc. As the graph shows, sample #903 is almost identical to aluminum.

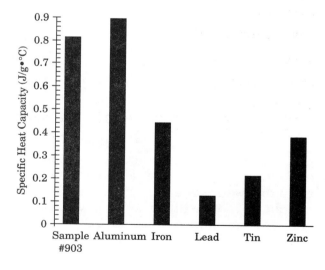

Sincerely,

Joe Doakes
Co-Director of Research
Superior California Analytical Laboratories

John Doe
Co-Director of Research

Notice the use of multiple trials. Both the data table and the graphs are in order. The explanation of the work and calculations done is clear and concise. More than one approach to measuring volume was attempted.

Evaluating Lab Work—Sample Reports, continued

Invoice Number: 4
Invoice Date: 10-10-94
Billed To: Maxwell Stone
                Operation Bird-in-the-Hand

EQUIPMENT	QUANTITY	RATES	AMOUNT
Metal sample	25 g	No charge	$0
Lab space	1	$15,000/day	$15,000
Disposal fee	25g	$2,000	$50,000
Balance, centigram	1	$5,000	$5,000
Beaker tongs	1 pair	$1,000	$1,000
Chemical handbook	1	$500	$500
25 mL graduated cylinder	1	$1,000	$1,000
400 ml beaker	2	$2,000	$4,000
Boiling chips	1 set	$500	$500
Glass stirring rod	1	$1,000	$1,000
Hot plate	1	$8,000	$8,000
Plastic foam cups	2	$1,000	$2,000
Scissors	1 pair	$500	$500
Test-tube holder	1	$500	$500
Large test tube	1	$1,000	$1,000
LEAP thermistor	1	$2,000	$2,000
Labor hours	10	$100/hr	$1,000

We appreciate your business!   Total $93,000

Evaluating Lab Work—Sample Reports, continued

CheMystery Labs
998 Marshall Rd.
Vacaville, CA
95687

October 17, 1994

Mr. Maxwell Stone
CIA Headquarters
Langley, Virginia

Dear Mr. Stone:

Our results indicate that the Drachmanian Separatists are the terrorist group responsible for the bombing of Crest Air flight 998. Tests on metal sample 955 lead to the conclusion that the device was made of iron.

Applying the equation of mass divided by volume of metal, we calculated the density of metal sample 955 as 5.95 g/cm³. The accepted value of iron is 9.87 g/cm³, which gave us a +24% error. Iron and tin were relatively close to the measured density of sample 955. The average of two trials of specific heat capacity was 0.51 J/g·°C Compared to the accepted value of iron, the percent error of our specific heat capacity is +15%. The most likely element was iron because it yielded the lower percent error for specific heat capacity. Therefore, tin was eliminated by specific heat testing.

To calculate the specific heat of metal sample 955, we used the technique of heating the sample and measuring the heat released with standard calorimetric techniques. The equation that is used for heat gained by the water in the calorimeter is the following.

$$\textbf{\textit{mass of }} H_2O \times \textbf{\textit{change in temp. of }} H_2O \times \textbf{\textit{4.180 J}}$$

Using this equation, the specific heat of the metal can be calculated with the following equation.

$$\frac{\textbf{\textit{heat gained by }} H_2O}{\textbf{\textit{mass of metal}} \times \textbf{\textit{change in temp. of metal}}}$$

The density can be calculated by using the following equation.

$$\frac{\textbf{\textit{mass of metal}}}{\textbf{\textit{volume of metal}}}$$

T22

Author's Example of Intermediate Level (4 points)

Density and Specific Heat Capacity Values for Sample 955

	Trial 1	Trial 2	Average
C_p	0.43 J/g•°C	0.59 J/g•°C	0.51 J/g•°C
Density	5.95 g/Cm^3		

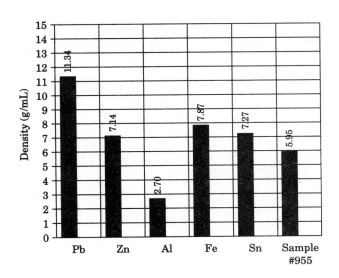
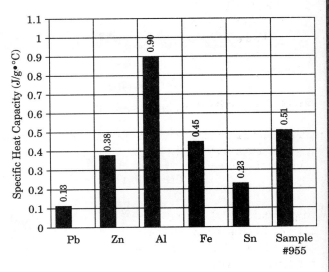

Sincerely,

Thomas Smith, Chemist
CheMystery Labs

Lucinda Wei, Chemist
CheMystery Labs

Although the data tables and graphs are in order, notice the small number of trials. The explanation of the work and calculations done is not very clear or concise. For example, the measurement of volume is not described. Although the students seem to have understood the lab, they are careless, as indicated by the value of "9.87 g/cm^3" given in the body of the letter for the density of iron. As the students' own graph shows, this should be "7.87 g/cm^3." According to the students' own graph of density measurements, the data indicate zinc was closer to the unknown than iron, but this is not discussed. The students report their percent error as a positive number instead of plus or minus.

Invoice

Ship To:

Bill To: **Maxwell Stone**
Operation Bird-in-the-Hand

Invoice Date	*Invoice Number*	*Purchase Order*	*Date Shipped*
October 10, 1994	2		

Quantity	Description	Unit Price	Extension
25 g	Metal sample	No charge	0.00
1	Lab space	15,000	15,000
25 g	Disposal fee	2,000	50,000
1	Balance	5,000	5,000
1 pair	Beaker tongs	1,000	1,000
1	Chemical handbook	500	500
1	Graduated cylinder (100 mL)	1,000	1,000
2	Beaker	2,000	4,000
1 set	Boiling chips	500	500
1	Glass stirring rod	1,000	1,000
1	Hot plate	8,000	8,000
2	Plastic foam cups	1,000	2,000
1 pair	Scissors	500	500
1	Test-tube holder	500	500
1	Test tube	1,000	1,000
1	Thermometer	2,000	2,000

Terms

Authorized by Lucinda Wei

Subtotal	$92,000
Sales tax	_____
Shipping	_____
TOTAL	$92,000

Author's Example of Beginning Level (2 points)

998 Marshall Rd.
Vacaville, CA
95687

October 19, 1994

Mr. Maxwell Stone:

The metal recovered at the crash site of flight 998 was sample 710,
which was proven to be iron. From this conclusion, the Drachmanian
Separatists are the guilty terrorist party. Specific heat capacity and
density were the properties used to identify the unknown sample as
iron. A calorimeter was used to determine the specific heat capacity of
the sample. The data from sample 710 was not exact, but is closest to
that of iron. Iron's exact specific heat capacity is 0.449 J/g·°C and
its exact density is 7.87 g/mL. The specific heat capacity of sample
710 was 0.4535 J/g·°C and it's density was 7.512 g/mL.

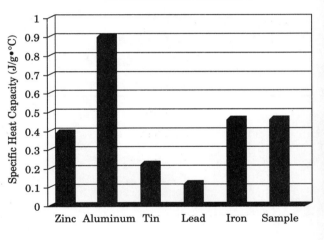

Trials	1	2	3	4
Specific Heat Capacity	0.5634	0.4535	0.5967	
Density	7.512	6.405	6.378	6.05

Sincerely,

Adriana N. Blanchette   Cy R. Damaso

> *Although an appropriate number of trials were run,
> instead of plotting averages, the students chose the
> trial that most closely fit the anticipated answer.
> There is no explanation of the experimental measure-
> ments and calculations done. There is no discussion
> of percent error for the data. Units have been left off
> the data table. The measurements in the data table
> have too many significant figures. In addition, the in-
> voice omits several pieces of equipment that were nec-
> essary, such as the balance and a test tube.*

Evaluating Lab Work—Sample Reports, continued

Author's Example of Beginning Level (2 points)

ADRIANA & CY
CHEMISTRY PERIOD 3

Date: 10/10/94

QUAN.	DESCRIPTION	UNIT COST	TOTAL COST
1	Metal sample	Free	Free
1	Lab space	15,000	15,000
25 g	Disposal fee	2,000	50,000
1 pair	Beaker tongs	1,000	1,000
1	Chemical handbook	500	500
100 mL	Graduated cylinder	1,000	1,000
2	400 mL beaker	2,000	4,000
1 set	Boiling chips	500	500
1	Glass stirring rod	1,000	1,000
1	Hot plate	8,000	8,000
2	Plastic foam cups	1,000	2,000
1 pair	Scissors	500	500
1	Test-tube holder	500	500
1	LEAP thermistor	$2,000	$2,000

		TOTAL COST	$86,000

Laboratory Safety

The statistical proof of this assertion is summarized by what is called the "Heinrick Accident Triangle." The base of the triangle represents 100 000 known and unknown unsafe conditions. Statistically, these 100 000 unsafe conditions will cause 10 000 close calls, which make up the next level of the triangle. Relating this statistic to the scale of your school's labs, if you had three close calls last year in your lab, you actually had 30 unsafe conditions. Or, if you had three recordable accidents, like a burn, a cut, or a spill that caused someone to slip and fall, you probably had 30 unreported close calls and 300 unsafe conditions, some of which you did not know about and therefore did not correct.

Heinrick Accident Triangle	Type of Occurrence	Hypothetical Triangle for Your Lab
1 0	Number of fatalities	?
1 0 0	Number of permanent disablements	?
1 0 0 0	Number of recordable incidents	3
1 0 0 0 0	Number of close calls	3 0
1 0 0 0 0 0	Number of known and unknown hazards	3 0 0

The direct cause of most accidents is an error caused by doing something unsafe or disobeying a known safety principle. ("I knew I was supposed to read the label.") But the underlying, indirect, and fundamental cause of every accident is not recognizing the importance of the precaution that, if it had been applied, would have prevented the error of commission or omission. ("I didn't know that paying attention to the label was that important.")

One method that can help students appreciate the importance of precautions is to use a "safety contract" that students read and sign, indicating they have read, understand, and will respect the necessary safety procedures, as well as any other written or verbal instructions that will be given in class. Consider making your own safety contract for your students with language specific to your lab situation. Typical contracts include points such as the following.

- Always wear personal protective equipment (goggles and lab aprons).
- Never wear contact lenses in the lab because chemical vapors can get caught behind them, causing eye damage.
- Read all lab exercises before you come to class. Follow all directions and safety precautions, and use only materials and equipment approved by your teacher.
- Remain alert and cautious at all times in the lab. Never leave experiments unattended.
- Do not wear heavy, dangling jewelry or bulky clothing.
- Only lab manuals and lab notebooks should be brought into the lab. Store backpacks, textbooks for other subjects, and other items elsewhere.
- No eating, drinking, or smoking is permitted in any science laboratory. **NEVER** bring food into the laboratory.
- **NEVER** taste chemicals. **NEVER** touch chemicals with your bare hands.
- Keep your head and clothing and other objects away from Bunsen burners. Light the burner only with a sparker. Be sure gas valves and hot plates are turned off before leaving the lab.
- Know the proper fire drill procedures and the locations of fire exits.
- Always clean and wipe dry all apparatus and work areas.
- Wash hands thoroughly with soap after cleanup at the end of a lab session.
- Know the location and operation of all safety equipment in the laboratory.
- Read the entire safety section of the lab manual and/or the textbook.
- Report all accidents or close calls to the teacher immediately, no matter how minor.
- **NEVER** work alone in the laboratory. Never work unless your teacher is present.

Eliminating unsafe conditions

Here is a list of several possible unrecognized causes of accidents and suggestions for accident prevention. Of course, the list is not complete; no such list can be. Use it to help you identify some of the known hazards in your laboratory. Once the potential causes of accidents are identified, you can then begin to work on their elimination.

Accidents waiting to happen	Precautions for prevention
Attitudes about safety	
• There is an attitude that precautions are optional if you are pressed for time.	• Practice zero-tolerance for unsafe conditions and practices. Nothing that can be accomplished in your lab is worth sacrificing someone's life or limb.
• There is no assessment of students' knowledge of and attitudes toward safety.	• Conduct frequent safety quizzes. Only students with perfect scores should be allowed to work in the lab.
• Neither you nor your supervisors have established a comprehensive safety plan.	• Include regular safety inspections and detailed record keeping in your safety plan.
• The recommendations of the most recent safety inspection or audit have not been incorporated into the safety program.	• Implement improvements to the safety plan immediately. Such improvements can protect you, your life, your limb, and your liability, but only if you implement them.
• The most recent safety inspection or audit was conducted more than six months ago.	• Conduct regular in-house safety and health inspections with an emphasis on improvement rather than guilt.
• You feel that things are safe enough already because you know that you and the other teachers will not do anything unsafe.	• Judge your safety considerations by whether they make the laboratory safe for everyone, including students, cleaning and building staff, and emergency workers.
Fires and other emergencies	
• Fire and other emergency drills are infrequent, and no records or measurements are made of the results of the drills.	• Always carry out critical reviews of fire or other emergency drills. Don't wait until an emergency to find the flaws in your plans.

Fires and other emergencies (contd)

- You have no idea what you'd do if an emergency happened in the middle of a lab exercise and the building had to be evacuated. You have never figured out how long it would take to secure the lab and any hazardous chemicals.

- Students are unsure what to do when the fire alarm or other alarm sounds.

- There are items stored in hallways along escape routes. Fire doors are propped open for ventilation.

- You know your escape route, but not any alternatives.

- The emergency phone numbers are kept only at the main switchboard.

- Fire extinguishers are placed at "dead ends" where they will not be in the way of people escaping.

- You have no idea how many fire extinguishers there are or when they were last charged. You have never used an extinguisher, and neither have the other teachers.

- Students believe that if a fire breaks out, they should run for the nearest fire extinguisher and use it to put out the fire themselves as quickly as possible.

- You and the other teachers nearby haven't had CPR or first-aid training in years.

- Have actions preplanned in case of an emergency (e.g., establish what devices should be turned off, which escape route to use, where to meet personnel outside the building, who is designated to count people at that meeting place, and who is authorized to re-enter).

- Inform everyone in the lab about all alarms and what to do if one sounds.

- Keep escape routes clear. Do not prop open fire doors or block them for any reason.

- Always plan alternate escape routes in case of unforeseen problems.

- Post current emergency phone numbers next to all phones.

- Place fire extinguishers near escape routes so that they can be used by those escaping.

- Regularly maintain fire extinguishers and train supervisory personnel in the proper use of extinguishers by extinguishing real fires.

- Instruct students (who should not use a fire extinguisher because they have not been trained in its use) that in case of a fire they should call a teacher (who should be trained).

- Get trained in CPR or first aid by your local chapter of the American Red Cross. Be sure to take frequent refresher courses.

Accident investigations

Accidents waiting to happen	Precautions for prevention
• The investigation conducted after a serious accident was superficial, did not identify any previously unrecognized causes. Safer procedures were not implemented.	• Record who worked with what, when, and for how long in order to allow meaningful retrospective studies of accidents.
• Accidents are analyzed only so that the person causing the accident can be disciplined.	• Analyze accidents to prevent repetition, not for any other reason.

Facilities and equipment

Accidents waiting to happen	Precautions for prevention
• Eyewash stations are present, but nobody knows anything about their specifications.	• Ensure that eyewash stations and safety showers meet the requirements of the ANSI standard (Z358.1).
• Eyewash stations are checked and cleaned only at the beginning of each school year.	• Run eyewash stations for 5 min every month to remove any bacteria or other organisms from pipes.
• No records are kept of routine checks and maintenance on the safety showers and eyewash stations.	• Test safety showers (measure flow in gallons per minute) and eyewash stations every six months and keep records.
• If the fan can be turned on, the lab hood is presumed to work well enough to keep everyone in the room safe.	• Ensure that laboratory ventilation and hood performance and use conform to the requirements of the ANSI standard (Z9.5).
• Chronic contamination of breathing air is dealt with by keeping windows propped open at all times.	• Engage an industrial hygienist to conduct the appropriate measurements when contamination of breathing air is suspected. Keep records of any such measurements.
• Spills are handled on a case-by-case basis with whatever materials happen to be on hand.	• Have the appropriate equipment and materials available for spill control; replace them before their expiration dates.
• Labs are opened in the morning and locked after school is out.	• Lock all laboratory rooms whenever a teacher is not present.
• Compressed gas cylinders are carried by hand and set up in a corner of the laboratory.	• Secure all compressed gas cylinders when in use, and transport them secured on a hand truck.
• Equipment with moving parts is used without any precautions.	• All moving belts and pulleys should have safety guards.

Do you really know how to operate an eyewash station?

Make sure you know exactly how—it could save someone's eyesight. Be certain to instruct **anyone working in the lab** in the proper use of the eyewash station, as outlined below.

- If your eyes are exposed to chemicals, walk immediately to an eyewash station, preferably **within 15 s of exposure.**
- Immediately start flushing your eyes; **continue for at least 15 min** while someone else calls a doctor.
- **Keep your eyes wide open;** with your fingers, lift your upper and lower lids away from the eyeball so that the flowing water washes across it.
- **Move your eyes continuously** up, down, left, right, across, and around. This will help the flowing water to get inside the eyeball socket and wash out any chemicals that could attack the eyeball from the rear.

Accidents waiting to happen	Precautions for prevention
Safety Wear	
• Instead of goggles, you prefer to wear safety glasses with hard plastic lenses. There is no face shield present to use when you are required to prepare solutions from concentrated acids.	• Wear eye protection in the laboratory at all times; use safety goggles, ANSI Type G or H. When circumstances require, wear face, neck, and ear protection also (face shield, ANSI Type N). For details see the ANSI standard (Z87.1).
• Students or teachers who must wear contact lenses use regular lab goggles.	• Require ANSI Type K cupped goggles for students or teachers who must wear contact lenses.
• Gloves are used over and over again to save money. They are discarded only after they begin to fall apart.	• Check gloves routinely for pinholes, tears, or rips. Because gloves cannot resist penetration by a chemical after being handled for long periods of time, they should be replaced before that time period expires.
• You wear old clothes in the lab instead of a laboratory coat or apron.	• Wear a laboratory coat or apron.
• You assume that your lab coat or apron offers the appropriate protection.	• Wear a long-sleeved shirt or blouse and slacks that extend to the ankles under the lab coat or apron.
• Your hair looks better when it is down, so you leave it that way when you work in the laboratory.	• Tie back or tuck in loose clothing (e.g., sleeves, full cut blouses, and neckties), long hair, and dangling jewelry.
• You allow students to wear sandals in the laboratory.	• Do not allow any footwear in the lab that does not cover feet completely; no open-toed shoes are allowed.
• Observers are allowed into the lab but are told to stay away from the lab benches.	• Keep a spare set of protective equipment on hand for visitors.
Work habits	
• New lab procedures are used with students without any planning or analysis.	• Analyze and try new lab procedures in advance to pinpoint potential hazards.
• You work alone during your preparation period to organize the day's labs.	• Never work alone in a science laboratory or storage area.

- Honor students do independent study alone in the laboratory.
- Although students do not eat or drink in the lab, you sometimes drink a soda or coffee in the storage area.
- You keep your lunch stored in the laboratory refrigerator because it's closer than the teacher's lounge.
- After work, you're in a hurry to get home, so you don't wash your hands.
- Sometimes you leave the room while you wait for a hot water bath to warm up.
- The storeroom is so crowded that you decide to keep some apparatus on the lab benches.

- The only precaution taken whenever volatile substances are used is to open the windows.
- When you work in the fume hood, you always put your head inside so that you can see what you are doing.
- Reagents are kept in the fume hood because you lack appropriate storage areas.
- Students who have finished their lab work take off their goggles when they begin writing up their results.
- You pipet by mouth because you know how it is done properly.
- No extra precautions are taken when handling liquid nitrogen.

- Never allow students to occupy a science laboratory unless a teacher is present.
- Never eat, drink, smoke, or chew gum or tobacco in a science laboratory or storage area.
- Do not store food or beverages in the laboratory environment.

- Always wash hands after work in a science laboratory and after spill cleanups.
- Never leave any heat sources (such as gas burners, hot plates, heating mantles, sand baths, etc.) unattended.
- Do not store reagents or apparatus on lab benches, and keep lab shelves organized. Never place reactive chemicals (in bottles, beakers, flasks, wash bottles, etc.) near the edges of a lab bench.
- Use a fume hood when working with volatile substances.

- Never lean into the fume hood.

- Do not use the fume hood as a storage area.

- Make sure protection is used not only by the lab worker but also by anyone working nearby.

- Never pipet by mouth.

- Tape all Dewar flasks.

Accidents waiting to happen	Precautions for prevention
• You always prepare your solutions from concentrated stock to save money.	• Reduce risks by ordering diluted instead of concentrated substances.
• You purchase plenty of chemicals to be sure there is enough.	• Purchase chemicals in class-size quantities, if at all possible.
• You purchase several year's worth of chemicals to save money.	• Do not purchase or have on hand more than one-year's supply of each chemical. Dispose of or use up any chemicals that are left over at the end of one year.
• When chemicals arrive, you unpack them and place them in your storage room without labeling them further.	• Accurately label all chemicals with the date of receipt and the initials of the person who unpacked that chemical.
• You are in a hurry to prepare solutions for a new lab, so you open the chemical and use it without reading the label.	• Never open a reagent package until the label has been read and completely understood.
• You never bother reading labels on chemicals because you've been using the chemicals for years and you already know all there is to know about them.	• Read each label to be sure it states the hazards and describes the precautions and first-aid procedures (when appropriate) that apply to the contents in case someone else has to deal with the chemical in an emergency.
• You never bother to read the Material Safety Data Sheet (MSDS) that comes with a chemical.	• Always read the MSDS for a chemical before using it. Follow the precautions described in the MSDS.
• You throw away the MSDSs after you read them because they just clutter up your storage area.	• File and organize the MSDS for each chemical in your lab so that they can be found easily in case of an emergency.
• You put a copy of the warnings from the concentrated acid label on a bottle of the diluted acid.	• Label diluted acids and bases with the hazards, precautions, and first aid procedures that apply specifically to them.
• Large reagent bottles of flammable chemicals are kept on the open shelves of the laboratory for days at a time.	• Store no more than a one-day supply of flammable liquids or solids on the open shelves of the laboratory. At the end of the day, store any such unused chemical in a flammable liquid storage cabinet.

T33

Accidents waiting to happen	Precautions for prevention

Purchasing and using chemicals (contd)

- Bottles of chemicals are kept unused on shelves in the lab during the semester. The main stockroom contains chemicals that haven't been used for many years.

- No extra precautions are taken when flammable liquids are dispensed from their containers.

- Do not leave bottles of chemicals unused on shelves in the lab for more than one week or unused in the main stockroom for more than one year.

- When transferring flammable liquids from bulk containers, ground the container, and before transferring to a smaller metal container, ground both containers.

Chemical storage

- Students are told to put their broken glass and solid chemical wastes in the trash can.

- Students retrieve chemicals and apparatus from stockrooms or storage areas.

- You're pretty sure you know what chemicals you have off the top of your head.

- For easy retrieval, you store your chemicals arranged in alphabetical order.

- You're not sure what chemicals are incompatible with what other chemicals.

- You keep mutually reactive chemicals stored by hazard class, but flammable liquids and solids are stored on the shelves of the storage area.

- Have separate containers for trash, for broken glass, and for different categories of hazardous chemical wastes.

- Lock all storage spaces; permit only teachers to enter.

- Keep an accurate and up-to-date inventory of chemicals on hand.

- Arrange chemicals by hazard class when storing them. (Alphabetical storage inevitably leads to adjacent positioning of incompatibles.)

- Use MSDSs to determine which chemicals are incompatible.

- Keep incompatible classes of mutually reactive chemicals (e.g., acids and bases, oxidizers and reducers, nitric acid and glacial acetic acid) separate from each other in the storage area. Store bulk quantities of flammable liquids and solids in a flammable materials storage cabinet or in a separate room specifically designed and designated to be used only for such storage.

T34

Accidents waiting to happen	Precautions for prevention
Chemical storage (contd)	

• Corrosives are kept above eye level, out of reach from anyone who is not authorized to be in the storeroom.	• Always store corrosive chemicals on shelves below eye level.
• Chemicals are kept on the floor of the stockroom on the days that they will be used so that they are easy to find.	• Never store chemicals or other materials on floors or in the aisles of the laboratory or storeroom, even for a few minutes.
• Chemicals are stored permanently on laboratory shelves because the storage room is too crowded.	• Return chemicals on the laboratory shelves to storage as soon as they are no longer needed.
• Because of a budget shortfall, stacked boxes are used instead of shelves.	• Equip chemical storage shelves with lips; never use stacked boxes instead of shelves.
• A second-hand refrigerator is used for lab storage.	• Use only an "explosion-proof" refrigerator for lab storage.
• Chemicals are stored without consideration of possible emergencies (fire, earthquake, flood, etc.) that could compound the hazards of the chemicals.	• Store chemicals that are incompatible with common firefighting media like water (such as alkali metals) or carbon dioxide (such as alkali and alkaline-earth metals) under conditions that eliminate the possibility of a reaction with water or carbon dioxide if it is necessary to fight a fire in the storage area.
• Batteries are stored without any extra precautions.	• Cover both terminals of dry cells and rechargeable batteries with insulating tape when storing them.

Disposal of Chemicals

Only a relatively small percentage of waste chemicals are classified as hazardous by EPA regulations. The EPA regulations are derived from two acts (as amended) passed by the Congress of the United States: RCRA (Resource Conservation and Recovery Act) and CERCLA (Comprehensive Environmental Response, Compensation, and Liability Act).

In addition, some states have enacted legislation governing the disposal of hazardous wastes that differs to some extent from the federal legislation. The disposal procedures described in this book have been designed to comply with the federal legislation as described in the EPA regulations.

In most cases the disposal procedures indicated in the teacher's edition will *probably* comply with your state's disposal requirements. However, to be sure of this, check with your state's environmental agency. If a particular disposal procedure does not comply with your state requirements, ask that office to assist you in devising a procedure that is in compliance.

The following general practices are recommended in addition to the specific instructions given in the Teacher's Notes.

- Except when otherwise specified in the disposal procedures, neutralize acidic and basic wastes with 1.0 M sodium hydroxide, NaOH, or 1.0 M sulfuric acid, H_2SO_4, added slowly while stirring.
- In dealing with a waste-disposal contractor, prepare a complete list of the chemicals you want to dispose of. Classify each chemical on your disposal list as hazardous or nonhazardous waste. Check with your local environmental agency office for the details of such classification.
- Unlabeled bottles are a special problem. They must be identified to the extent that they can be classified as a hazardous or nonhazardous waste. Some landfills will analyze a mystery bottle for a fee if it is shipped to the landfill in a separate package, is labeled as a sample, and includes instructions to analyze the contents sufficiently to allow proper disposal.

Electrical Safety

Although none of the labs in this manual require electrical equipment, several include options for the use of microcomputer-based laboratory equipment, pH meters, or other equipment. The following safety precautions to avoid electric shocks must be observed any time electrical equipment is present in the lab.

- Each electrical socket in the laboratory must be a three-hole socket and must be protected with a GFI (ground-fault interrupter) circuit.
- Check the polarity of all circuits before use with a polarity tester from an electronics supply store. Repair any incorrectly wired sockets.
- Use only electrical equipment equipped with a three-wire cord and three-prong plug.
- Be sure all electrical equipment is turned off before it is plugged into a socket. Turn off electrical equipment before it is unplugged.
- Wiring hookups should be made or altered only when apparatus is disconnected from the power source and the power switch is turned off.

- Do not let electrical cords dangle from work stations; dangling cords are a tripping and shock hazard.
- Do not use electrical equipment with frayed or twisted cords.
- The area under and around electrical equipment should be dry; cords should not lie in puddles of spilled liquid.
- Hands should be dry when using electrical equipment.
- Do not use electrical equipment powered by 110–115 V alternating current for "conductivity" demonstrations or for any other use in which bare wires are exposed, even if the current is connected to a lower voltage AC or DC connection.
- Use dry cells or ni-cad rechargeable batteries as direct current sources. Do not use automobile storage batteries or AC-to-DC converters; these two sources of DC current can present serious shock hazards.

Prepared by Jay A. Young, Consultant, Chemical Health and Safety, Silver Spring, Maryland

References

1. Budavari, Susan, ed., *The Merck Index,* 11th ed., Merck and Co., Rahway, NJ (1989).
Use this reference for reliable general information about a chemical instead of using a less reliable chemical dictionary or one of the many books with the words "Dangerous Properties" as part of the title.

2. Council Committee of Chemical Safety, *Safety in Academic Chemistry Laboratories,* 5th ed., American Chemical Society, Washington, DC (1995).
This is the authority if you wish a brief treatment. Single copies are free. For a more extensive treatment see Young, *Improving Safety in the Chemical Laboratory,* below.

3. "Fire Protection for Laboratories Using Chemicals" (also known as "NFPA-45"), National Fire Protection Association, Batterymarch Park, Quincy, MA (current edition).
This is the national safety code for laboratory fire protection and prevention.

4. Gerlovich, J. A., et al., *School Science Safety,* Flinn Scientific, Inc., Batavia, IL (1984).
This is a practical guide in two volumes for teachers.

5. Pipitone, D. A., and D. Hedberg, "Safe Chemical Storage," *Journal of Chemical Education,* 59 (1982), A159.
A discussion on the proper storage of chemicals. Also see Pipitone, D. A., *Safe Storage of Laboratory Chemicals,* 2nd ed., Wiley-Interscience, New York, NY (1991) and the Flinn Scientific Catalog, Flinn Scientific, Inc., Batavia, IL (current edition).

6. "Practice for Occupational and Educational Eye and Face Protection, Z87.1," "Emergency Eyewash and Shower Equipment, Z358.1," "American National Standard for Laboratory Ventilation, Z9.5," American National Standards Institute, New York, NY (current editions).
These standards are recognized worldwide as reliable. Safety goggles and face shields that meet Z87.1 requirements are marked *Z87.*

7. Reese, K. M., *Health and Safety Guidelines for Chemistry Teachers,* American Chemical Society, Washington, DC (1980).
This source contains several useful tips on enhancing safety in your laboratory.

8. *Standard First Aid and Personal Safety,* American Red Cross.
Your local American Red Cross chapter probably gives a short course in first aid. This book summarizes the content of that course.

9. Young, Jay A., "Risk Assessment and Hazard Evaluation for Undergraduate Laboratory Experiments," *Journal of Chemical Education,* 59 (1982), A265.
This provides some suggestions on teaching your students how to practice safety in the lab.

10. Young, Jay A., ed., *Improving Safety in the Chemical Laboratory: A Practical Guide,* 2nd ed., Wiley-Interscience, New York, NY (1991).

Laboratory Program Supplies

Chemicals

This list shows the chemicals needed for 15 lab groups to perform all of the laboratory exercises in this lab manual, both *Explorations* and *Investigations*. Many of these items will be consumed during the course of the laboratory. Instructions for reclaiming or recycling the other reagents are given in the Teacher's Notes for each lab.

For solutions, the list includes both descriptions of the concentrations and volumes required for the laboratory exercise as well as the amount of pure or concentrated reagent that should be ordered. Many chemical suppliers provide a variety of ready-made solutions. Ordering chemicals already diluted may be slightly more expensive, but it is much safer. A list of chemical suppliers is provided on page T51 of the annotated teacher's edition of the textbook.

Item	Order quantity (for 15 lab groups)	Lab(s)
Acetic acid, ~5%, 750 mL	38 mL glacial acetic acid	Inv. 13-3
$Al_2(SO_4)_3$, 0.5 M, 3.375 L	1.0 kg $Al_2(SO_4)_3 \cdot 14H_2O$	Exp. 15-2 Inv. 15-2
Aluminum metal sample, 25 g	60 (1.5 kg total)	Exp. 9-1 Inv. 9-1
Aluminum strip, 1 cm × 8 cm	45	Exp. 15-2 Inv. 15-2
Boiling chips	100	Exp. 9-1 Inv. 9-1
$Ca(C_2H_3O_2)_2 \cdot 2H_2O$ (calcium acetate)	300.0 g	Inv. 12-2
$CaCl_2$	100.0 g	Exp. 12-2
CaO (lime)	7.5 g	Inv. 13-2
Chalk, antiglare	375 g	Exp. 7-1
Charcoal, activated	120 g	Inv. 2-1
Coconut oil	150 g	Inv. 9-2
Copper metal sample, 25 g	45 (1.125 kg total)	Exp. 9-1
Copper strip, 1 cm × 8 cm	45	Exp. 15-2 Inv. 15-2
Copper wire, 10 cm lengths	15 (1.5 m total)	Exp. 15-1
$CuCl_2 \cdot 2H_2O$	375 g	Inv. 12-1
$CuSO_4$, 225 g $CuSO_4 \cdot 5H_2O$, 0.5 M, 3.375 L	648 g $CuSO_4 \cdot 5H_2O$	Exp. 15-2 Inv. 12-1, 15-2
Dissolved oxygen test kits*	60 tests	Exp. 11-2

Item	Order quantity (for 15 lab groups)	Lab(s)
Distilled water	as needed	throughout
Dry ice	400 g	Exp. 10-2
Ethanol (denatured)	910 mL	Exp. 7-1, 15-1 Inv. 6-1, 13-3
$FeCl_3$	48 g $FeCl_3 \cdot 6H_2O$	Exp. 15-1
Filter paper, 12 cm diameter	195 pieces	Exp. 2-1 7-1, 11-1 Inv. 11-1
$H_2O_2(aq)$, 3%	2.65 L	Exp. 14-1 Inv. 14-1
$H_2SO_4(aq)$, conc., for teacher use only	5 mL	Inv. 13-2
HCl, 1.0 M, 1.5 L	130 mL conc. HCl	Exp. 15-1 Inv. 13-1
Ice	as needed	throughout
Iron filings	40 g	Exp. 2-1
Iron metal sample, 25 g	60 (1.125 kg total)	Exp. 9-1 Inv. 9-1
Iron strip, 1 cm × 8 cm	15	Exp. 15-1
Isopropanol	1.2 L	Exp. 7-1, 11-1 Inv. 11-1
Lubricant (mineral oil or glycerin)	5 mL	Inv. 10-1
Magnesium strip, 1 cm × 8 cm	30	Inv. 15-2
Methane from gas jet	as needed	Exp. 10-2

Item	Order quantity (for 15 lab groups)	Lab(s)
$MgSO_4$, 0.5 M, 225 mL	28 g $MgSO_4 \cdot 7H_2O$	Inv. 15-2
MnO_2	15 g	Inv. 14-1
$Na_2CO_3 \cdot 10H_2O$	45 g	Inv. 6-1
$Na_2S_2O_3 \cdot 5H_2O$	675 g	Exp. 9-2, 9-4
$Na_3C_6H_5O_7 \cdot 2H_2O$ (sodium citrate)	7.5 g	Exp. 12-1
$NaC_2H_3O_2$ (sodium acetate)	450 g	Exp. 9-4
NaCl	710 g	Exp. 2-1, 12-1, 12-2 Inv. 6-1, 12-2
$NaHCO_3$	436 g	Exp. 1-1 Inv. 12-2, 13-2
NaI	22.5 g	Exp. 14-1 Inv. 14-1
NaOH, 1.0 M, 1.65 L	66.0 g NaOH pellets	Inv. 13-1, 13-3
$NH_3(aq)$, (ammonia water), conc.	100 mL	Exp. 11-3
$(NH_4)_2HC_6H_5O_7$ (ammonium citrate)	48 g	Exp. 15-1
NH_4Cl	29 g NH_4Cl	Exp. 15-1
pH paper, narrow range	135 pieces	Inv. 13-1, 13-2, 13-3
Phenolphthalein solution	1.0 g phenolphthalein	Inv. 13-3
Salicylic acid (HOC_6H_4COOH)	45 g	Inv. 6-1

Item	Order quantity (for 15 lab groups)	Lab(s)
$SnCl_2$, 0.5 M, 225 mL	26 g $SnCl_2 \cdot 2H_2O$	Inv. 15-2
Sucrose	145 g	Exp. 12-2 Inv. 6-1
Tin strip, 1 cm $\times$ 8 cm	30	Inv. 15-2
Weighing paper	1 package, 500 sheets	throughout
Wooden splint	30	Exp. 2-1 Inv. 2-1
Zinc strip, 1 cm $\times$ 8 cm	60	Exp. 15-1, 15-2 Inv. 15-2
$ZnSO_4$, 60 g $ZnSO_4 \cdot 7H_2O$, 0.5 M, 3.375 L	545 g $ZnSO_4 \cdot 7H_2O$	Exp. 15-1, 15-2 Inv. 15-2

* Alternatively, a dissolved oxygen meter or a LEAP System dissolved oxygen probe (see *Optional Materials and Equipment*) may be used. Do not use dissolved oxygen test kits that make use of the Winkler method or other titrimetric methods such as LaMotte's test kits because of the dangerous chemicals and the difficulties related to disposal. A simple and less expensive dissolved oxygen test kit is available from scientific suppliers such as Ward's Natural Science Establishment (1-800-962-2660). The kit contains small reagent-filled ampuls sealed in a vacuum. When the tip is broken off and dipped into a sample, the sample is drawn inside to react with the reagents, after which it can be compared to standards provided in the kit. When used properly, the user will not come into contact with any of the reagents. For Exploration 11-2, each lab group needs four test ampuls.

| Ward's | 21 W 9001 | 10 test ampuls and a set of standard ampuls |
| | 21 W 9002 | refill kit with 10 additional test ampuls |

Equipment

This list shows the equipment and laboratory apparatus needed for 15 lab groups to perform all of the laboratory exercises in this laboratory manual, both *Explorations* and *Investigations*. All of these items are reusable or are used to make equipment that can be reused each time the laboratory exercise is performed.

In some cases, individual *Explorations* or *Investigations* can be performed more rapidly if more glassware is available so that students do not need to stop to clean it until the entire laboratory exercise is completed.

In other cases, students performing the open-ended *Investigations* may request more equipment to perform their procedure than is actually necessary. Some students may even request equipment that is not listed. These students' proposed procedures must be examined carefully. Provided they are safe, you may decide whether or not to grant their requests.

Item	Order quantity (for 15 lab groups)	Lab(s)
Balance, centigram	3 or more	throughout
Beaker tongs	15	throughout
Beaker, 50 mL	105	Exp. 6-2 Inv. 6-2
Beaker, 150 mL	75	throughout
Beaker, 250 mL	90	Exp. 7-1, 11-3, 12-2, 15-2 Inv. 12-2, 13-1, 15-2
Beaker, 400 mL	90	throughout
Beaker, 600 mL	45	Exp. 9-2, 9-4, 11-2 Inv. 13-2
Bunsen burner*	15	throughout, especially Inv. 6-1
Buret clamp, double	15	Exp. 14-1 Inv. 13-1, 13-3, 14-1
Buret tube	30	Inv. 13-1, 13-3
Chemical storage refrigerator	as needed	throughout
Clamp, test-tube	30	Inv. 10-1
Conductivity tester, battery-operated	15	Inv. 6-1
Crucible tongs	15	Exp. 7-1, 10-2, 12-1 Inv. 12-1, 12-2

Item	Order quantity (for 15 lab groups)	Lab(s)
Evaporating dish	45	Exp. 7-1, 12-1 Inv. 12-1, 12-2
Flask, Erlenmeyer, 125 mL	45	Exp. 14-1 Inv. 2-1, 13-3, 14-1
Forceps	15	Exp. 2-1, 9-2
Fume hood	as needed	throughout
Glass bottle, 2 L	15	Exp. 10-2
Glass funnel†	15	Exp. 2-1, 7-1
Glass stirring rod	45	throughout
Graduated cylinder, 10 mL	15	Exp. 12-1 Inv. 12-1
Graduated cylinder, 25 mL	15	Exp. 6-2, 12-2 Inv. 2-1, 6-2
Graduated cylinder, 100 mL	15	throughout
Graduated cylinder, 250 mL	15	Exp. 10-2
Hot plate*	30	throughout, especially Inv. 12-1
Leveling bulb	15	Exp. 14-1 Inv. 14-1
Magnets	30	Exp. 2-1

Item	Order quantity (for 15 lab groups)	Lab(s)
Medicine dropper	30	Exp. 10-2, 14-1 Inv. 14-1
Microscale reaction strip, 8-well	15	Exp. 2-1
Mohr buret, 100 mL	15	Exp. 14-1 Inv. 14-1
Mortar and pestle	15	Exp. 7-1
Petri dish with lid	15	Exp. 2-1, 11-1 Inv. 11-1
Pin, straight	15	Inv. 6-2
Pinch clamp	15	Exp. 10-2
Pipet, microscale, graduated	16	Inv. 2-1
Pipet, microscale, jumbo-sized	45	Exp. 2-1 Inv. 2-1
Pipet, microscale, thin-stem	195	Exp. 2-1, 6-2, 11-3 Inv. 2-1, 6-2, 13-1, 13-3
Pipet, microscale, wide-stem	15	Exp. 11-3
Pneumatic trough	15	Exp. 10-2, 14-1 Inv. 14-1
Porous cup‡	15	Exp. 15-2 Inv. 15-2
Ring clamp	45 for Exp. 7-1, 15 elsewhere	throughout
Ring stand	45 for Exp. 7-1, 15 elsewhere	throughout
Rubber stopper, for test tubes	120	Exp. 2-1 Inv. 6-1
Rubber stopper, one-hole, for Erlenmeyer flask	15	Exp. 7-1, 14-1 Inv. 2-1, 14-1

Item	Order quantity (for 15 lab groups)	Lab(s)
Rubber stopper, one-hole, for Mohr buret	15	Exp. 14-1 Inv. 14-1
Rubber stopper, one-hole, no. 6	15	Exp. 10-2
Rubber stopper, one-hole, split	15	Inv. 10-1
Rubber tubing, for Bunsen burner*	15 pieces	throughout
Rubber tubing, 20 cm pieces	30 (6.0 m total)	Exp. 14-1 Inv. 14-1
Rubber tubing, 50 cm pieces	15 (7.5 m total)	Exp. 10-2
Ruler, metric	15	Exp. 6-2, 9-2, 10-1, 11-3 Inv. 6-2
Salt bridge‡	45	Exp. 15-2 Inv. 15-2
Scissors	15	Exp. 9-1, 11-3 Inv. 9-1
Sealed syringe, 10 cm³	15	Inv. 10-1
Spatula	15	Exp. 12-1 Inv. 12-1, 13-2
Striker	15	throughout
Syringe, sealed§	15	Exp. 10-1
Test tube, large, Pyrex (at least 25 × 150 mm)	120	throughout
Test tube, medium, Pyrex	15	Exp. 9-2
Test tube, small, Pyrex	105	Exp. 6-2, 9-4 Inv. 6-2
Test-tube clamp, three-fingered	15	Exp. 9-1, 9-2 Inv. 9-1

Item	Order quantity (for 15 lab groups)	Lab(s)
Test-tube holder	15	throughout
Test-tube rack	15	throughout
Thermometer clamp	15	Exp. 6-2, 9-3 Inv. 6-2
Thermometer, nonmercury	45 for Exp. 7-1, 30 for Exp. 9-2, and 15 elsewhere	throughout
Voltmeter	15	Exp. 15-2 Inv. 15-2
Wash bottle	15	Inv. 2-1, 6-1, 13-1, 13-3
Wax pencil	15	Exp. 6-2, 15-1 Inv. 6-2
Wire, small gauge	400 cm	Exp. 9-2, 9-4

Item	Order quantity (for 15 lab groups)	Lab(s)
Wire gauze, with ceramic center*	45 for Exp. 7-1, 30 for Exp. 12-1, and 15 elsewhere	throughout
Wire, with alligator clips	60 pieces	Exp. 15-1, 15-2 Inv. 15-2

* A Bunsen burner must be used for Investigation 6-1. Two hot plates per lab group are recommended for Investigation 12-1. Otherwise, either hot plates or Bunsen burners may be used for the remainder of the laboratory experiences.

† Alternatively, a vacuum filtration arrangement with a Büchner funnel and a vacuum filter flask can be used, as indicated in the *Optional Materials and Equipment* section.

‡ Either salt bridges or porous cups (but not both) are necessary for the electrochemical cells constructed in Exploration 15-2 and Investigation 15-2.

§ Instead of a sealed syringe, a Boyle's law apparatus can be used for Exploration 10-1, as indicated in the *Optional Materials and Equipment* section.

Miscellaneous Materials

The laboratory experiences in *Holt Chemistry: Visualizing Matter* emphasize more use of consumer products than do many traditional lab manuals. This list shows the miscellaneous materials needed for 15 lab groups to perform all of the laboratory exercises in this laboratory manual, both *Explorations* and *Investigations*. Most of these items will be consumed during the course of the laboratory experiences.

Item	Order quantity (for 15 lab groups)	Lab(s)
Aluminum foil	1 roll	Exp. 2-1
Battery, 6 V lantern	15	Exp. 15-1
Beet, chopped*	250 mL	Inv. 13-1
Blackberry, chopped*	250 mL	Inv. 13-1
Blueberry, chopped*	250 mL	Inv. 13-1
Carpet thread, 15 cm piece	15	Exp. 10-1
Cellophane	1 roll	Exp. 2-1
Chemical handbook or reference	at least 1	Exp. 9-2, 9-4 Inv. 9-1
Chemical supply catalog	at least 1	Exp. 9-4
Coffee, ground	5 mL	Inv. 2-1
Corrugated cardboard	40 cm × 40 cm	Exp. 9-4
Cotton ball	45	Exp. 2-1
Cranberry, chopped*	250 mL	Inv. 13-1
Emery cloth	15	Exp. 15-2 Inv. 15-2
Garlic powder	5 mL	Inv. 2-1
Jars, with screw-on lids	60	Exp. 11-2
Label, stick-on	15	Exp. 15-1

Item	Order quantity (for 15 lab groups)	Lab(s)
Milk carton, cardboard, for teacher use	2	Exp. 12-2
Mineral oil	10 mL	Inv. 2-1
Oil, SAE-10	450 mL	Exp. 6-2 Inv. 6-2
Oil, SAE-20	450 mL	Exp. 6-2 Inv. 6-2
Oil, SAE-30	450 mL	Exp. 6-2 Inv. 6-2
Oil, SAE-40	450 mL	Exp. 6-2 Inv. 6-2
Oil, SAE-50	450 mL	Exp. 6-2 Inv. 6-2
Oil, SAE-60	450 mL	Exp. 6-2 Inv. 6-2
Paper clip	1 box of 100	Exp. 2-1, 11-1 Inv. 11-1
Paper towel	as needed	throughout
Peach skin, chopped*	250 mL	Inv. 13-1
Pear skin, chopped*	250 mL	Inv. 13-1
Pencil	15	Exp. 9-4, 11-1 Inv. 11-1
Pen, black	at least six different types	Exp. 11-1 Inv. 11-1
Pin, straight	15	Exp. 6-2

Item	Order quantity (for 15 lab groups)	Lab(s)
Plastic bag, 1 L	15	Exp. 1-1, 10-2
Plastic cup, clear	30	Exp. 1-1
Plastic foam cup	30	Exp. 9-1, 9-3 9-4 Inv. 9-1, 12-2
Plastic fork	15	Exp. 2-1
Plastic spoon	15	Exp. 2-1 Inv. 2-1
Plastic straw	15	Exp. 2-1, 10-2
Plastic tray	15	Inv. 2-1
Plastic washtub	15	Exp. 9-2, 9-4
Poppy seeds	40 g	Exp. 2-1
Radish skin, chopped*	250 mL	Inv. 13-1
Red apple skin, chopped*	250 mL	Inv. 13-1
Red cabbage, chopped*	250 mL	Inv. 13-1
Red cherry, chopped*	250 mL	Inv. 13-1
Red onion, chopped*	250 mL	Inv. 13-1
Removable Post-It note	60	Exp. 10-1
Rhubarb, chopped*	250 mL	Inv. 13-1
Rubber band	15	Exp. 10-2
Safety match (for teacher use only)	45	Exp. 7-1

Item	Order quantity (for 15 lab groups)	Lab(s)
Sand, coarse-fine mixture	265 g	Exp. 2-1 Inv. 2-1
Solid fuel, such as Sterno	15 pieces	Exp. 7-1
Stackable object (books or blocks of wood), ~500 g each	60	Exp. 10-1
Steel wool	15 pieces	Exp. 15-1
Strawberry, chopped*	250 mL	Inv. 13-1
Tallow (or lard)	150 g	Inv. 9-2
Tape	1 roll	Exp. 2-1
Tissue paper	2 boxes	Exp. 2-1 Inv. 2-1
Tomato skin, chopped*	250 mL	Inv. 13-1
Turnip skin, chopped*	250 mL	Inv. 13-1
Twist tie	30	Exp. 1-1
Vegetable shortening	150 g	Inv. 9-2
Vinegar (acetic acid solution)	6.0 L	Exp. 1-1, 7-1
Yeast	15 g	Inv. 14-1
Yellow onion, chopped*	250 mL	Inv. 13-1

* These materials represent some of the possibilities for use in Investigation 13-1, which explores the use of natural, edible substances as pH indicators. Not all of these materials need to be used, and other substances can be substituted.

To help fit your situation and resources, the laboratory experiences in *Holt Chemistry: Visualizing Matter* also contain alternative procedures for different types of equipment, including the LEAP System microcomputer-based laboratory probes. In every case, a procedure using more traditional equipment is also given in the student portion of the procedure.

Item	Order quantity (for 15 lab groups)	Lab(s)
Aspirator, for spigot (vacuum filtration)	15	Exp. 7-1
Barometer	1	Exp. 10-1
Beaker tongs	1	Exp. 15-1
Boyle's law apparatus	15	Exp. 10-1
Büchner funnel	15	Exp. 7-1
Calorimeter	15	Exp. 9-3
Dissolved oxygen meter	1 or more	Exp. 11-2
Drying oven	1	Exp. 15-1
Glass beads	150	Exp. 9-1
LEAP conductivity probe	15	Inv. 6-1
LEAP dissolved oxygen probe	15	Exp. 11-2
LEAP pH probe	15	Inv. 13-1, 13-2, 13-3
LEAP System*	45 inputs for Exp. 7-1, 30 inputs for Exp. 9-2 and 11-2 and Inv. 13-1 and 15 inputs elsewhere	throughout

Item	Order quantity (for 15 lab groups)	Lab(s)
LEAP thermistor probe	45 for Exp. 7-1, 30 for Exp. 9-2, and 15 elsewhere	throughout
LEAP voltage probe	15	Exp. 15-2 Inv. 15-2
Magnetic stirrer	15	Inv. 13-2
Magnets	15	Inv. 9-1
pH meter	at least 3	Inv. 13-1, 13-2, 13-3
Spot plates, microscale	15	Inv. 13-1
Stopwatch	15	throughout
Tea infusers	15	Exp. 9-1
Vacuum flask (sidearm flask), and tubing	15	Exp. 7-1

* To order the LEAP System and probes, contact Quantum Technology. Quantum has set up a variety of packages for users of *Holt Chemistry: Visualizing Matter.*

Quantum Technology
30153 Arena Drive
Evergreen, CO 80439

Sales and Technical Support: (303) 674-9651
Fax: (303) 674-6763

Conservation of Mass

Objectives

Students will
- use appropriate lab safety procedures.
- demonstrate proficiency in detecting a chemical reaction.
- measure masses of reactants and products using a laboratory balance.
- measure volumes of liquids using a graduated cylinder.
- resolve chemical discrepancies.
- design chemical experiments.
- relate observations of a chemical reaction to the law of conservation of mass.

Planning

Recommended Time

1 lab period

Materials

(for each lab group)
- Baking soda ($NaHCO_3$), 15 g
- Vinegar (acetic acid, CH_3COOH, solution), 300 mL
- 400 mL beaker
- 100 mL graduated cylinder
- Balance
- Clear plastic cups (or 150 mL beakers), 2
- Large plastic bag
- Twist ties, 2
- Weighing papers, 2

Solution/Material Preparation

1. Use vinegar and baking soda purchased through a supply company or at a grocery store instead of diluted glacial acetic acid.
2. Provide fairly large plastic bags, such as the 1 qt size available at most stores.

Student Orientation

Techniques to Demonstrate

The less you demonstrate for the students and the more they figure out on their own, the better. It may be necessary to show them that you can use a twist tie to keep the baking soda isolated in one corner of the bag so that vinegar can be poured in without them reacting immediately.

Pre-Lab Discussion

From other courses, students will probably be familiar with the concepts of chemical

Required Precautions

- Goggles and a lab apron must be worn at all times.
- Read all safety cautions, and discuss them with your students.
- In case of an acid or base spill, first dilute with water. Then, mop up the spill with wet cloths or a wet cloth mop while wearing disposable plastic gloves. Designate separate cloths or mops for acid and base spills.

change and conservation of mass. This lab can be used as a discrepant event that challenges what "everybody knows" about these topics. To explain what has actually happened and come up with a procedure that truly tests conservation of mass, students will have to think about what they observed and imagine what is going on at the level of individual particles. Students may ask to be shown the equation for the reaction. If they need hints, the following leading questions may help them understand why mass did not appear to be conserved in Part 1.
- What evidence was there of a chemical reaction?
- Were the products of the reaction solids, liquids, or gases?
- How well do the containers you used hold liquids? How well do they hold gases?
- How can you use the plastic bag and twist ties to be certain that the reaction takes place in a container that holds liquids and gases?

Proposed Procedure for Part 2

The second half of the lab is designed to be an open-ended inquiry. Students need to come up with a way to make sure that the reaction system is truly closed. What follows is one possibility. Although student answers will vary, be sure that they will ensure a closed reaction system.

Measure out 4–5 g of baking soda. Pour it into the corner of the bag. Using a twist tie, separate that corner from the rest of the bag. Add about 50 mL of vinegar to the bag. Use another twist tie to seal the top of the bag so that the vinegar cannot leak out and the bag is airtight. Place the bag in the 400 mL beaker. Measure the mass of the bag, the beaker, and the reactants. Remove the twist tie that kept the baking soda separate from the vinegar, and allow the two to mix. Replace the bag in the beaker. As the reaction occurs, the bag will inflate, but do not remove the twist tie that keeps the bag closed and airtight. After the reaction is complete, measure the mass of the bag, beaker, and products.

Note that because the bag inflates, a mass loss of as much as 0.5–0.7 g will be found even if the bag is perfectly sealed, due to the effect of the buoyant force of the air displaced. If you wish to avoid this error, one way of reducing it is to inflate the bag with air to the approximate volume it is expected to be filled by the CO_2 and seal it with a twist tie prior to measuring its mass the first time. Thus, the volume change and the effects of the buoyant force that it induces will be minimized. Then remove the twist tie, deflate the bag completely, re-seal it with the twist tie, and then allow the reactants to mix.

Post-Lab

Disposal

All of the solutions and chemicals used in this investigation may be washed down the sink with an excess of water.

Sample Data

	Initial mass (g)	Final mass (g)	Change in mass (g)
Part 1	122.3	119.8	2.5
Part 2	73.4	73.2	0.2

Answers to

Analysis and Interpretation

1. The bubbling and fizzing of the reactants was evidence of a chemical change. It indicated that a gaseous product was being formed. The baking soda reactant seemed to disappear eventually.

Conclusions

2. Although there appeared to be a loss of mass in Part 1, the container did not adequately confine the gaseous product, which was not accounted for when the final mass was measured.

3. The system in Part 1 consisted of two cups and the reactant. The system was open, so the CO_2 product was able to escape. In Part 2, the system was closed, and the gaseous product was confined. As a result, there was no change in the mass.

4. Students' answers describing instructions for the display at the mall will vary. Be sure answers are safe, include carefully planned procedures, and demonstrate an understanding of the need to confine mass. Proposed instructions should roughly follow the proposed procedure given earlier in these Teacher's Notes.

5. Students' explanations of how the scientific method was applied will vary, but they should recognize that observations were made and data were collected to test the hypothesis that mass is conserved. When the results did not match the hypothesis, the experiment was scrutinized for potential sources of error and tried in a new way. Then, more observations and data were collected. The results were interpreted to indicate support for the hypothesis that mass is conserved.

6. Students' explanations of how a theme from the text relates to the procedure will vary. Some possible answers are the following.

   Systems and interactions The results of the interaction could be measured only in a completely closed system, as in Part 2.

   Equilibrium and change There was evidence of a chemical change when the reactants were combined.

Macroscopic observations and micromodels The fact that the mass measurement changed in Part 1 and not in Part 2 reflects the fact that individual atoms were neither created nor destroyed.

Extensions

1. The burning of a log produces products such as smoke, CO_2, and water vapor, all of which escape as the log burns. The ashes are only one of the products of the reaction.
2. Students' answers will vary. Encyclopedias and books about chemical history should contain information about Lavoisier's life and work.
3. Students' answers will vary. Be certain each suggestion is a valid conversion of energy. One possible answer is that chemical energy in the gasoline is converted into heat energy as it is combusted. The heat energy is converted into energy of motion, which turns the wheels and moves the car forward.
4. Students should refer to a reference work such as the *CRC Handbook of Chemistry and Physics, Lange's Handbook of Chemistry,* or *The Merck Index.* There is some slight disagreement among these sources.

	Molar mass, g/mol	Density, at 25°C, g/mL	Melting point, °C	Boiling point, °C
$NaHCO_3$	84.01	2.159	decomposes 270	
CH_3COOH	60.06	1.049	16.6	117.9
$NaCH_3COO$	82.04	1.528	324	decomposes
H_2O	18.02	1.00	0.0	100.0
CO_2	44.01	1.81×10^{-3}		−78.5 sublimes

5. Student's answers will vary. They should indicate that acetic acid is mildly caustic and that carbon dioxide is an asphyxiant. However, the situation here does not present a great hazard because small quantities are being used. Students should recommend that safety goggles be worn.

Additional Notes

Separation of Mixtures

EXPLORATION
2-1
Technique
Builder

Objectives

Students will
- use appropriate lab safety procedures.
- identify chemical and physical properties of substances.
- relate knowledge of properties to the task of separating mixtures.
- identify methods of separating the items.
- design and implement their own procedure.
- separate the components of a mixture.
- analyze the success of methods of purifying mixtures.

Planning

Recommended Time

2 lab periods (evaporation time may be necessary)

Materials

*(for each lab group)**
- Distilled water
- Sample of mixture, 10 g
- 8-well microchemistry strip
- Aluminum foil
- Cellophane
- Cotton balls
- Filter paper
- Forceps
- Glass funnel
- Magnets
- Microfunnel
- Paper clips
- Paper towels
- Petri dish
- Pipets
- Plastic forks
- Plastic spoons
- Plastic straws
- Rubber stoppers
- Tape
- Test-tube holder
- Test-tube rack
- Test tubes
- Tissue paper
- Wood splints

* Note that not all of this equipment is useful for separating the mixture. It is still recommended that you have all of it available so that students are required to think through what they will do.

Required Precautions

- Goggles and a lab apron must be worn at all times to provide protection for your eyes and clothing.
- Read all safety cautions, and discuss them with your students.

Solution/Material Preparation

1. To make a microfunnel, cut the stem and the top of the bulb off a jumbo pipet, as shown in the illustration. Students can use these funnels to fill the wells in their 8-well microchemistry strip.

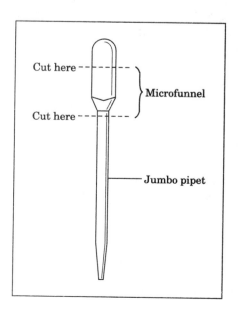

Cut here — Microfunnel
Cut here
Jumbo pipet

2. To make the mixture, take each of the four components and mix them. About 100 g at a time can be made if 25 g of each component are poured together in a zippered, resealable plastic bag. After shaking, the mixture may be dispensed to the students in stoppered test tubes, each containing 10 g of the mixture.

3. For an additional challenge, you can provide smaller amounts of the mixture. However, students are unlikely to be successful with less than 2 g.

4. Space should be available for air-drying samples overnight.

Student Orientation

Techniques to Demonstrate

Be certain students understand how to perform a filtration. Detailed instructions for setting up the apparatus are included as part of Exploration 1B in the textbook.

Pre-Lab Discussion

Be certain students realize the importance of thinking through their procedures before they begin. Point out that once they make a mistake with their sample, it is likely to set them back even further than if they take time to think about it first.

Post-Lab

Sample Separation Procedure

(Note: students' procedures may vary.)

Spread the mixture on a paper towel. Wrap a magnet in a piece of cellophane. Pass the magnet over and through the mixture, shaking loose non-iron particles. Repeat this technique to remove as much iron as possible. Unwrap the magnet, keeping the iron inside the cellophane. The iron is now separated and recovered.

Place the sample in a petri dish. Add some distilled water. The poppy seeds should float to the top. Remove them carefully with a plastic spoon. The poppy seeds are now separated and recovered.

Stir the mixture until the salt is entirely dissolved. Fold a piece of filter paper in quarters, and place it in a glass funnel. Transfer the remaining mixture from the petri dish to the filter. Catch the filtrate with a test tube. Add small amounts of water to the petri dish and repeat until all of the sand has been washed onto the filter paper. Remove the filter paper from the funnel,

and let it dry overnight. The sand is now separated and recovered.

Pour the saltwater filtrate into the petri dish. Rinse the tube with distilled water to be sure all the salt is transferred back to the petri dish. Allow the water to evaporate overnight. Then use a plastic spoon to scrape out as many salt crystals as possible. The salt is now separated and recovered.

Disposal

Put out five disposal containers for the students, for iron, poppy seeds, sand, salt, and the unseparated mixture, respectively. The materials may be reused next year, even if the students have not done a particularly good job of separation, because you will be mixing the substances again the next time the Exploration is performed.

Answers to

Analysis and Interpretation

1. Students' answers will vary. See the sample procedure above for one possible solution. As justifications for their estimations of success, they may mention the presence (or absence of) impurities in the separated components and the amount recovered or the time it took to complete the separation.

Conclusions

2. For most students, the order of the procedure will be very important. For example, if the filtration is the first step, when you are finished, you will still have the wet sand-iron-poppy-seed mixture to sort out.

3. Students' suggestions for changing the procedure will vary. Be certain that their suggestions include safe, thorough procedures.

4. Students' suggestions for additional equipment will vary. One likely suggestion is the use of a hot plate or drying oven to speed evaporation steps. Pay attention to their suggestions so that you can add this equipment to the list the next time you have students perform this Exploration.

5. Students' identification of properties will probably vary but may include: iron is attracted by a magnet, salt dissolves in water, poppy seeds float in water, and sand sinks in water.

6. Students' answers will vary. In much of science, purity is often more important than the speed of the process, but not always. Accept either answer, provided that the reasoning for it is complete and thorough.

7. Students' suggestions for scoring will vary. A possible scoring method could be to make a determination of component's purity on a scale of 1 to 10, add the values together, and divide by the time it took to complete the procedure. Then these scores could be compared for different groups. Be certain that students apply their scoring plan to the results for the rest of the class.

Extensions

1. Students' suggestions for determining purity will vary but may include ideas such as comparing the densities of their samples to the densities of standard samples of the pure components or picking through their samples and checking for visible impurities.

2. Possible answers:
   a. Lead filings and iron filings can be separated with a magnet.
   b. Sand and gravel could be separated by a screen or sifter.
   c. Sand and finely ground plastic foam could be separated by adding water; the foam will float, and the sand will sink.
   d. Sugar will dissolve in some substances that salt will not.
   e. Because they have different boiling points, alcohol and water can be separated through distillation.
   f. Because they have different boiling points, nitrogen and oxygen can be separated through distillation.

3. Students' presentations about petroleum and coal products will vary. Be sure that students identify the stages of the process in which mixture separation techniques are used.

4. Students' presentations of the water treatment process will vary. Be sure students indicate what impurities are removed at each step.

Additional Notes

Separation of Mixtures—Tanker Truck Spill

INVESTIGATION 2-1

Problem Solving

Objectives

Students will
- use appropriate lab safety procedures.
- relate properties of components in a mixture to the techniques used to separate them.
- demonstrate proficiency in separating mixtures of liquids.
- observe the properties of the mixture throughout the separation.
- calibrate a wooden splint to serve as a measure of liquid volume.
- design and implement their own procedure.

Planning

Recommended Time

1 lab period

Prerequisite

Exploration 2-1

Materials

(for each lab group)
- Charcoal (activated), 8 g
- Distilled water
- Foul water samples, each about 3 mL
- Sand (coarse-fine mixture), 15 g
- 25 mL graduated cylinder
- Long-stemmed microfunnels, 2 or more
- Pipet bulb without stem
- Pipet (thin-stemmed)
- Plastic spoon
- Plastic tray
- Strainer
- Test-tube holder
- Test-tube rack
- Test tubes, 6
- Tissue paper or cotton
- Wash bottle
- Watch glass
- Wire for cleaning pipets, small gauge (10–12)
- Wooden splint

For Teacher Use
- 125 mL Erlenmeyer flask

- One-hole rubber stopper
- Graduated pipet

Estimated cost of materials: $58 000

Required Precautions

- Goggles and a lab apron must be worn at all times to provide protection for eyes and clothing.
- Read all safety cautions, and discuss them with your students.

Solution/Material Preparation

1. Use the following recipe to make 1.0 L of the foul water. In an easy-to-clean or disposable container, combine 800 mL of tap water, 1 tablespoon of sifted, ground coffee, and 1 tablespoon of sifted garlic powder. Shake or stir well. Add 200 mL of mineral oil and shake or stir.

2. To dispense consistently mixed and sized samples of the foul water, pour some of it in a 125 mL Erlenmeyer flask. Slide a graduated plastic (beral) pipet through the one-hole rubber stopper. Put the stopper and pipet on the flask, as shown in the illustration. Shake vigorously. Squeeze the bulb of the pipet. Slide the

pipet out, and dispense a sample of the foul water into the student's test tube. Repeat the procedure.

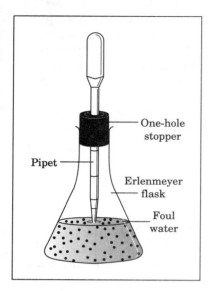

Pipet
One-hole stopper
Erlenmeyer flask
Foul water

3. For the charcoal, purchase aquarium grade activated charcoal. Use a mortar and pestle to crush it until it has a granular consistency. The charcoal and the sand can be reused, as indicated in the disposal section.

4. To make the bulbs for separating the oil and water, cut off all but 4 or 5 mm of the stem from a graduated or thin-stemmed pipet, as shown in the illustration. This bulb functions similarly to the classic large-scale separatory funnels used in other labs. Instead of using a stopcock to release the more dense phase, pressure on the bulb forces the more dense phase out the bottom of the bulb. These bulbs can be reused every time this Investigation is performed.

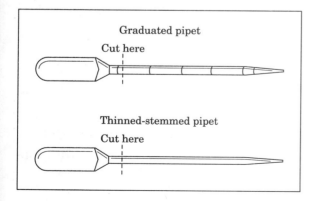

Graduated pipet
Cut here

Thinned-stemmed pipet
Cut here

5. To make long-stemmed microfunnels, cut the top off of the bulb of a jumbo pipet, as indicated in the illustration. Each group should use at least two microfunnels, but you may want to have more on hand for students who may ask for additional ones. These funnels can be reused every time this Investigation is performed.

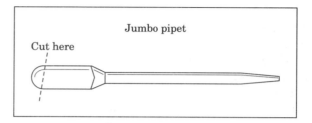

Jumbo pipet
Cut here

6. To save time, consider calibrating the wooden splints yourself, following the directions given in the Student Orientation section. Once one has been calibrated, you can use it as a template for the others, provided all students are issued test tubes of the same size.

7. Either a wad of tissue paper or a piece of a cotton ball can be used to plug the bottom of the microfunnel.

Student Orientation

Techniques to Demonstrate

Be sure to show students how to use each of the pieces of equipment made from pipets. They will need to be shown that the best way to put the liquid into the bulb for the oil-water-sediment separation is to pour it onto a watch glass and use a thin-stemmed pipet to suck the water up. Then, insert the stem of the pipet into the bulb for separation, and gently squeeze to dispense the sample.

Make sure students understand that they must turn the bulb with the opening down for the settling step that separates the sediment, water, and oil into separate layers. Gentle, steady pressure must be applied when squeezing out the layers. If students suddenly release too much pressure, air may be sucked into the bulb, causing mixing of the layers.

Remind students that they will need to use a wad of tissue paper to prevent the sand or charcoal from falling out the bottom of their long-stemmed microfunnels. To speed up the charcoal-filtering step, they can place their thumb over the top rim of the funnel and press down to increase the pressure.

Be sure students understand the need for a calibration step. Add 1.0 mL of water to one of their test tubes, and place it in the rack. Hold the wooden splint next to the test tube, and draw a line on the splint level with the bottom of the meniscus in the test tube. Repeat the process until at least 5.0 mL can be measured with the splint. Because the volumes in this lab are so small and the measurements must be made after each step, measuring volume with a graduated cylinder is not very practical.

Begin by discussing the results and procedure used in the Exploration, especially any errors in lab technique that occured. To help students, you may provide some of the following leading questions for them to consider as they make their plans.

- How many components are in your mixture?
- What properties are different for the components?
- Which method (removing layers, filtering with sand, filtering with charcoal) is likely to remove the very smallest impurities?
- What disadvantages are there to using the method that removes the smallest particles first?

Students should identify at least the following properties for the components as ones that can be used for the separation: density, solubility, and particle size. Be sure plans for separation are reasonable and take into account that the most effective approach starts with the techniques that remove the largest impurities and finishes with the techniques that remove the smallest ones. Sand filters should be layered coarse, fine, coarse for best results. If time permits, you may allow atudents to discover this themselves.

Proposed Procedure

Measure the volume of the foul water sample using a calibrated wooden splint, and record the volume and the sample's properties in the data table. Pour the sample onto an evaporating dish and pick it up using a thin-stemmed pipet. Using the thin-stemmed pipet, deliver the sample into the bulb. Turn the bulb mouth-side down, and let it stand undistrubed until two liquid layers have formed. Carefully squeeze the first drop or two of sediments back into one test tube. Then, deliver the water layer to another clean test tube, being careful not to allow any of the top oily layer to drip out. Finally squeeze out the oil layer into the first test tube. Measure the volume of the water

	Before	After oil-water step	After sand	After charcoal
Color	brown	light brown	none	none
Clarity	cloudy	slightly cloudy	slightly cloudy	clear
Odor	bad	mild	mild	none
Oil present?	yes	none	none	no
Solids present?	yes	some	none	no
Volume (mL)	3.0	2.3	2.1	1.9

$$\frac{1.9 \text{ mL } H_2O \text{ recovered}}{3.0 \text{ mL in foul water sample}} \times 100 = 63\%$$

(Students' percentages should be less than 80%.)

Students should observe that the charcoal step takes longest, and that it is most efficient to remove most of the impurities through other means. That way, the volume of mixture that must pass through the charcoal is minimized.

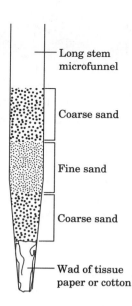

Long stem
microfunnel

Coarse sand

Fine sand

Coarse sand

Wad of tissue
paper or cotton

layer in the second test tube, and record the volume and properties of the sample in the data table.

Place a wad of tissue paper or cotton in the stem of a long-stemmed microfunnel. Add some coarse and fine sand in layers (Coarse, fine, coarse is the best arrangement). Place the microfunnel in a third test tube, and carefully pour the water sample into it. When the filtration appears to be complete, remove the microfunnel. Measure the volume of the filtered water sample, and

record the volume and properties of the sample in the data table.

Place another wad of tissue paper in the bottom of another microfunnel, and add charcoal. Place the funnel into the fourth test tube. Carefully pour the water from the third test tube into the funnel. Measure the final volume of the purified water, and record the volume and properties in the data table. Repeat the filtration process using sand and charcoal filters in tandem. After removing sediment and oil layers, deposit water sample into sand filter. Allow the filtrate to drip onto the charcoal filter. Collect the purified water from beneath the charcoal filter with a test tube.

Post-Lab

Disposal

Set out three disposal containers—one for sand, one for charcoal, and one for the oily layer and sediment. The purified, filtered water may be poured down the drain. The sand may be washed, dried, and used again. To reuse the charcoal, wash it, and dry it for several hours in the oven. The oven step helps re-activate the charcoal. Oily paper towels used to soak up any oil left on test tubes and other equipment can be a fire hazard, so dispose of them promptly.

Additional Notes

Covalent and Ionic Bonding—Ceramics Fixative

INVESTIGATION 6-1

Problem Solving

Objectives

Students will
- use appropriate lab safety procedures.
- relate the properties of ionic and covalent substances to properties they observe and measure.
- heat substances in a test tube to observe melting.
- mix substances with ethanol and water to determine their solubilities.
- design and implement their own procedure.

Planning

Recommended Time
1 lab period

Prerequisite
None

Materials

(for each lab group)
- Distilled water, 24 mL
- Ethanol, 24 mL
- Salicylic acid, 3.0 g
- Sodium carbonate, 3.0 g
- Sodium chloride, 3.0 g
- Sucrose, 3.0 g
- 150 mL beakers, 3
- 100 mL graduated cylinder
- Balance
- Bunsen burner
- Conductivity tester, battery-operated
- Gas tubing for Bunsen burner
- Glass stirring rod
- Rubber stoppers (for test tubes), 8
- Striker
- Test-tube holder
- Test-tube rack
- Test tubes, 8
- Wash bottle
- Weighing paper, 12 pieces

Optional Equipment
- Conductivity probe—LEAP

Estimated Cost: $114 000–$118 000

Required Precautions

- Goggles and a lab apron must be worn at all times to provide protection for eyes and clothing.
- Tie back long hair and loose clothing.
- Read all safety cautions, and discuss them with your students.
- Remind students that heated objects can be hot enough to burn even if they look cool. Students should always use test-tube holders or tongs when handling any lab equipment that has been heated.
- Be certain that the ethanol is kept in a closed bottle within an operating fume hood. Do not dispense any ethanol until all students have completed the melting point tests, and all burners have been turned off and have cooled down. Make only 300 mL of ethanol available at a time. Students should replace the lid as soon as they have taken what they need.

continued on next page

- No burners, flames, hot plates, or other heat sources should be in use in the lab during the times that students are working with ethanol.
- NEVER use conductivity meters that rely on 110 V or 115 V alternating current. NEVER use automobile storage batteries or AC to DC converters as power sources.
- If the LEAP system with conductivity probe is used, the precautions listed in the teacher's notes on page T37 must be followed to avoid electric shock.

Solution/Material Preparation

1. Inexpensive battery-powered conductivity meters are available from many scientific suppliers. **NEVER use conductivity meters that rely on 110 V or 115 V alternating current. NEVER use automobile storage batteries or AC to DC converters as power sources.**
2. You can also make your own inexpensive conductivity apparatus with the following materials.
   Battery, 9 V, with battery clips
   Electrician's tape
   Film canister with lid
   LED (light-emitting diode)
   Resistor, 1.0 kΩ (0.25 W)

Straw or tubing, 3.5 cm
Nail or icepick
Wire, 22 gauge, solid, 30 cm
Hot-glue gun (optional)

Cut the wire into a 10 cm piece and a 20 cm piece. Strip about 1.5 cm of insulation from the ends of the hookup wires and battery leads. Cut the resistor leads down to half their length. Using a nail or icepick, poke two holes on the top of the canister cap about as far apart as the LED leads. Insert one lead through each hole, and spread them apart slightly.

Connect the circuit as shown in the illustration. **Be sure the RED battery clip lead is connected to the longer LED lead.** The black battery clip lead should connect to the resistor, and the other end of the resistor should connect to the 10 cm wire. The shorter LED lead should be connected to the 20 cm wire. Bend the leads and tape them against the inside of the lid so that they are not touching each other. Wrap tape around the resistor and all connections.

Near the bottom of the film canister, poke two holes about 1 cm apart. Insert the free ends of the 10 and 20 cm wires through these holes. Connect the battery, and place it in the canister, snapping the lid shut. Adjust the protruding hookup wires so that they extend about 5 cm. Drops of hot-melt glue may be placed around the LED wires and hookup wires to make the canister water-proof. Wrap black electrician's tape around the straw, and place it over the LED to provide a

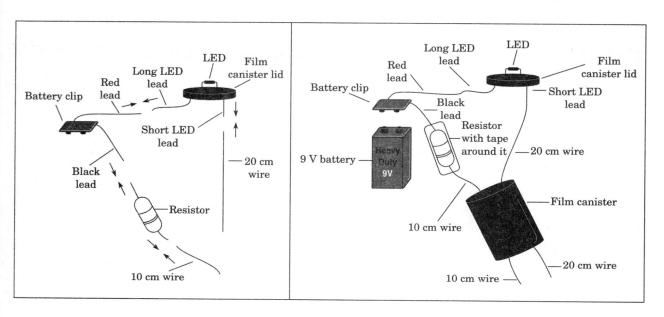

"dark tunnel" through which the LED may be viewed for testing weak electrolytes.

Student Orientation

Techniques to Demonstrate

Review the proper method for heating a test tube with students. Show students how to use the LED conductivity tester. Remind students to rinse the leads on the conductivity tester with distilled water before and after each test.

Pre-Lab Discussion

Students may be uncertain about performing an Investigation without doing an Exploration first, but all of the concepts necessary for success in this lab are covered in Chapters 5 and 6. It may help reassure students if you relate the properties of ionic and covalent substances to the nature of each bond type. Everything else they need to know about procedure, such as how to test for dissolving in water and ethanol and how to perform the melting-point test, is provided in the documents they have received. If you will be using the LEAP conductivity probe, explain to students that greater values of conductivity mean the substance conducts electricity.

Tips for Evaluating the Pre-Lab Requirements

Be certain that students have accounted for the proper order in their procedure. As specified in the Proposed Procedure, all work with Bunsen burners for melting must be completed before the solubility portion of the

lab begins. Students should explain exactly what they will be attempting to observe and why for each of the tests.

Proposed Procedure

Heat 0.5 g of each substance in a test tube, using a test-tube holder, over a Bunsen burner for exactly 1.0 min. Observe and record whether melting occurs. Repeat for all substances.

Place 1.0 g of each substance in its own test tube. Add 5.0 mL of water to each tube, stopper, and shake. Observe and record which dissolve. Repeat the test with new 1.0 g samples of each substance and ethanol.

For those compounds that dissolved in water, pour the solutions into a 150 mL beaker. Use the conductivity probe to determine whether they conduct electricity.

Post-Lab

Disposal

Set out and label six disposal containers. Four are for students to dispose of excess amounts of the four substances tested. The other two are for water solutions and ethanol solutions, respectively. Reuse any excess amounts of the substances tested the next time this Investigation is performed. Filter the water solutions and the ethanol solutions to remove any solids. These may be placed in the trash after they have dried. The water solutions may be poured down the drain. After dilution with at least 10 times its volume with water, the ethanol solution may also be poured down the drain.

Sample Data and Analysis

Compounds	Melts?	Dissolves in H_2O?	Dissolves in C_2H_5OH?	Conducts?
Salicylic acid	yes	no*	yes	no
Sodium carbonate	no	yes	no	yes
Sodium chloride	no	yes	no*	yes
Sucrose	reacts	yes	yes	some

* Although students will report that no salicylic acid dissolved in water and no sodium chloride dissolved in ethanol, actually these compounds are very slightly soluble.

Students should indicate that sodium carbonate and sodium chloride are the compounds that most closely match the needs outlined in the letter. Some students may suggest that sodium chloride is a better choice because sodium carbonate readily absorbs moisture from the air to form a hydrate, as was discussed in Ch. 5.

Viscosity of Liquids

EXPLORATION
6-2
Technique Builder

Objectives

Students will

- use appropriate lab safety procedures.
- demonstrate proficiency in comparing the viscosity of various liquids under identical test conditions.
- construct a small viscosimeter.
- measure flow time of various single-weight oils using a stopwatch or clock with second hand.
- measure the mass using a laboratory balance.
- measure volume using a graduated cylinder.
- calculate density from measurements of mass and volume.
- calculate the relative viscosity of the oils.
- graph experimental data.
- compare viscosities and densities to determine the SAE rating of each oil.

Planning

Recommended Time

1 lab period

Materials

(for each lab group)
- Distilled water
- Ice
- Oil samples, 10 mL, 6 (SAE-10, -20, -30, -40, -50, -60)
- 50 mL beakers, 7
- 400 mL beakers, 2
- 10 mL graduated cylinder
- Pin, straight
- Pipets, thin-stem, 7
- Ruler, metric
- Stopwatch or clock with second hand
- Test-tube holder
- Test-tube rack
- Test tubes, small, 7
- Wax pencil

Bunsen burner option
- Bunsen burner
- Gas tubing
- Ring stand and ring
- Striker
- Wire gauze with ceramic center

Hot plate option
- Hot plate

Thermometer option
- Thermometer, nonmercury
- Thermometer clamp

Required Precautions

- Goggles and a lab apron must be worn at all times to provide protection for the eyes and clothing.
- Tie back long hair and loose clothing.
- Read all safety cautions, and discuss them with your students.
- Remind students to use a test-tube holder when removing test tubes from the warm-water bath.
- The oils used in this Exploration should not be heated above 60°C.
- If a hot plate or the LEAP System with thermistor probe is used, the precautions listed in the teacher's notes on page T37 must be followed to avoid electric shock.

LEAP System option
- LEAP System
- LEAP thermistor probe

Solution/Material Preparation

1. For the oil samples, collect the grades of oil indicated in the following table, and label them with the letters *A–F*. It does not matter which type of oil gets labeled with which letter, provided you keep a

record. The sample data and calculations shown in these teacher's notes are based on the following table.

Oil type	Letter
SAE-10	C
SAE-20	A
SAE-30	D
SAE-40	F
SAE-50	B
SAE-60	E

2. The oil samples should be reused from class to class to avoid the need to dispose of them.

3. Once the pipet viscosimeters have been made and labeled, they can be reused from class to class. For best results, use only one grade of oil in each pipet viscosimeter.

4. It may help students complete the lab in a single lab period if you provide two sets of samples in test tubes. This way, students can be cooling and heating the different sets at the same time.

Student Orientation

Techniques to Demonstrate

Be sure to demonstrate exactly how to prepare the pipet viscosimeter. Show students how to control the flow by covering the hole with a finger. It may take several trials before students have a consistent technique that allows measurements to be made. If students have trouble completing the lab in the time allotted, reduce the number of trials run at each temperature.

Sample Data and Analysis

Sample	Beaker mass (g)	Total mass (g)	Volume (mL)	Trial 1—cool (s)	Trial 2—cool (s)	Trial 3—cool (s)
A	47.06	51.51	5.0	31.1	31.2	31.3
B	46.81	51.45	5.0	148.5	148.4	148.6
C	47.27	51.63	5.0	20.3	20.2	20.1
D	48.04	52.55	5.0	43.5	43.7	43.6
E	47.53	52.22	5.0	190.5	190.4	190.3
F	47.60	52.15	5.0	74.5	74.6	74.4
H_2O	46.92	51.92	5.0	1.6	1.6	1.6

Sample	Trial 1—room temp. (s)	Trial 2—room temp. (s)	Trial 3—room temp. (s)	Trial 1—warm (s)	Trial 2—warm (s)	Trial 3—warm (s)
A	14.7	14.6	14.5	7.3	7.2	7.1
B	51.3	51.4	51.2	18.3	18.4	18.2
C	10.5	10.3	10.4	5.5	5.3	5.4
D	20.9	21.1	21.0	9.8	9.9	9.7
E	73.6	73.7	73.5	25.5	25.3	25.4
F	31.7	31.6	31.8	11.9	11.7	11.8
H_2O	1.6	1.5	1.6	1.5	1.6	1.5

cool temperature: 4°C
room temperature: 20°C
warm temperature: 40°C

Make certain that the concepts of inter-molecular bonding and hydrocarbon chain length discussed in Chapter 6 of the text-book are understood. Point out that the larger the molecule, the stronger the London forces can become.

Later, the technique used here of moving the finger away from a hole to allow liquid to flow out of the viscosimeter can be used as an object lesson to help students understand the role of atmospheric pressure. When a finger is held over the hole, the atmospheric pressure outside the pipet holds the fluid inside the stem.

Post-Lab

Disposal

Set out twelve disposal containers, one for each of the six types of oil labeled A through F and for each of the six different pipets labeled A through F. Students in successive periods can reuse the labeled beakers, test tubes, and pipets. There is no need for a full-scale cleanup of all equipment until the end of the class day. The oil can be reused class after class and year after year. If the oil is cleaned up with paper towels, they can be disposed of only in a landfill designated for hazardous waste.

Answers to

Analysis and Interpretation

1. A: $\dfrac{4.45 \text{ g}}{5.0 \text{ mL}} = 0.89 \text{ g/mL}$

    B: $\dfrac{4.64 \text{ g}}{5.0 \text{ mL}} = 0.93 \text{ g/mL}$

    C: $\dfrac{4.36 \text{ g}}{5.0 \text{ mL}} = 0.87 \text{ g/mL}$

    D: $\dfrac{4.51 \text{ g}}{5.0 \text{ mL}} = 0.90 \text{ g/mL}$

    E: $\dfrac{4.69 \text{ g}}{5.0 \text{ mL}} = 0.94 \text{ g/mL}$

    F: $\dfrac{4.55 \text{ g}}{5.0 \text{ mL}} = 0.91 \text{ g/mL}$

    H_2O: $\dfrac{5.00 \text{ g}}{5.0 \text{ mL}} = 1.0 \text{ g/mL}$

2.

Sample	Avg. time (s) at 4°C	Avg. time (s) at 20°C	Avg. time (s) at 40°C
A	31.2	14.6	7.2
B	148.5	51.3	18.3
C	20.2	10.4	5.4
D	43.6	21.0	9.8
E	190.4	73.6	25.4
F	74.5	31.7	11.8
H_2O	1.6	1.6	1.5

3. relative viscosity$_A$ =
    $$\frac{0.89 \text{ g/mL} \times 14.6 \text{ s} \times 1.002 \text{ cp}}{1.0 \text{ g/mL} \times 1.6 \text{ s}} =$$
    8.1 cp

    relative viscosity$_B$ =
    $$\frac{0.93 \text{ g/mL} \times 51.3 \text{ s} \times 1.002 \text{ cp}}{1.0 \text{ g/mL} \times 1.6 \text{ s}} =$$
    30. cp

    relative viscosity$_C$ =
    $$\frac{0.87 \text{ g/mL} \times 10.4 \text{ s} \times 1.002 \text{ cp}}{1.0 \text{ g/mL} \times 1.6 \text{ s}} =$$
    5.7 cp

    relative viscosity$_D$ =
    $$\frac{0.90 \text{ g/mL} \times 21.0 \text{ s} \times 1.002 \text{ cp}}{1.0 \text{ g/mL} \times 1.6 \text{ s}} =$$
    12 cp

    relative viscosity$_E$ =
    $$\frac{0.94 \text{ g/mL} \times 73.6 \text{ s} \times 1.002 \text{ cp}}{1.0 \text{ g/mL} \times 1.6 \text{ s}} =$$
    43 cp

    relative viscosity$_F$ =
    $$\frac{0.91 \text{ g/mL} \times 31.7 \text{ s} \times 1.002 \text{ cp}}{1.0 \text{ g/mL} \times 1.6 \text{ s}} =$$
    18 cp

Conclusions

4. If the oils were labeled as in the table in the Solution/Material Preparation section, students should identify the following matches. Otherwise, refer to your notes about which oil was labeled with which letter.

Letter	Rel. viscosity (cp)	Oil type
A	8.1	SAE-20
B	30.	SAE-50
C	5.7	SAE-10
D	12	SAE-30
E	43	SAE-60
F	18	SAE-40

5.

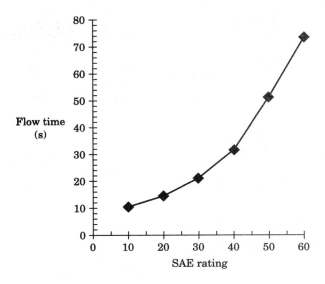

6.

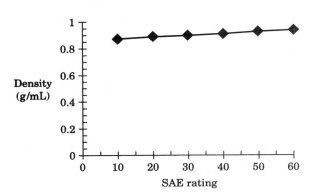

7.

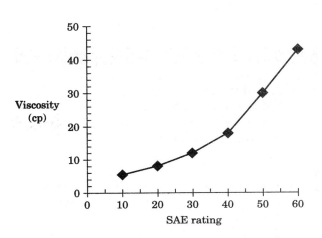

8.

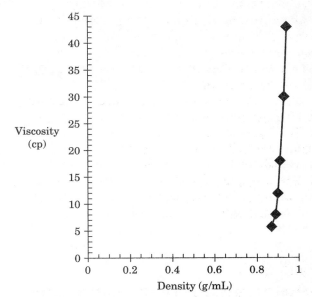

9. All of the oil samples are more viscous at low temperatures and less viscous at high temperatures.

10. The densities of the various oils do not vary by much. However, some students may notice that as the density increases, so does the viscosity.

11. The lower the SAE rating, the lower the viscosity of the oil.

12. More viscous fluids take more time to flow.

Extensions

1. Given the patterns detected so far, SAE-35 should have flow times between those of SAE-30 and SAE-40. Answers should be near 59.0 s for 4°C, 26.4 s for 20°C, and 10.8 s for 40°C.

2. If Malcolm heats the bottle of pancake syrup, the syrup will be less viscous and should flow more quickly.

3. Student answers will vary. Be certain students realize that the different properties of the different grades of oil are a result of the different combinations of ingredients used for each grade.

Viscosity— New Lubricants

INVESTIGATION 6-2

Problem Solving

Objectives

Students will
- use appropriate lab safety procedures.
- relate the properties of a mixture of oils to the properties of the components of the mixture.
- predict the composition of a multigrade oil that will have certain properties.
- prepare mixtures of oils that will have these properties and compositions.
- construct a small viscosimeter.
- demonstrate proficiency in comparing the viscosity of various liquids under identical test conditions.
- measure time using a stopwatch or a clock with a second hand.
- measure the mass using a laboratory balance.
- measure volume using a graduated cylinder.
- calculate density from measurements of mass and volume.
- calculate the relative viscosity of the oils.
- graph experimental data.
- compare viscosities and densities to determine the SAE rating of each oil.
- design and implement their own procedure.

Planning

Recommended Time

1 lab period

Prerequisite

Exploration 6-2

Materials

(for each lab group)
- Distilled water
- Ice
- Oil samples, SAE-10 to SAE-60, 40 mL of each type used
- 50 mL beakers, 4
- 400 mL beakers, 2
- 25 mL graduated cylinder
- Glass stirring rod
- Pin, straight
- Pipets, thin-stem, 4
- Ruler, metric
- Stopwatch or clock with second hand
- Test-tube holder

continued on next page

Required Precautions

- Goggles and a lab apron must be worn at all times to provide protection for the eyes and clothing.
- Tie back long hair and loose clothing.
- Read all safety cautions, and discuss them with your students.
- Remind students to use a test-tube holder when removing test tubes from the warm-water bath.
- The oils used in this Exploration should not be heated above 60°C.
- If a hot plate or the LEAP System with thermistor probe is used, the precautions listed in the teacher's notes on page T37 must be followed to avoid electric shock.

- Test-tube rack
- Test tubes, 6
- Wax pencil

Bunsen burner option
- Bunsen burner
- Gas tubing
- Ring stand and ring
- Striker
- Wire gauze with ceramic center

Hot plate option
- Hot plate

Thermometer option
- Thermometer, nonmercury
- Thermometer clamp

LEAP System option
- LEAP System
- LEAP thermistor probe

1. Most students should have planned to use a 1:1 mixture of SAE-20 and SAE-30 oil for one of the multigrade oils and a 1:1 mixture of SAE-10 and SAE-20 for the other. They should only need 20 mL of SAE-20 and 10 mL of SAE-10 and SAE-30. However, to avoid giving away clues, it would be best to have at least 20 mL of the other grades of oil available for every lab group.
2. The viscosimeters used in the Exploration may be reused here. For best results, use only pipets that had been filled with oils whose SAE ratings are similar to those in the mixtures.
3. The pure oil samples should be reused from class to class in order to avoid the need to dispose of them.
4. If motor oils are unavailable to you, similar work can be done by mixing various vegetable oils, mineral oil, and glycerin, but some adaptation will be necessary.

Student Orientation

Techniques to Demonstrate

Discuss any errors in technique observed during the Exploration. Because the testing procedures are the same, there should be few questions about the mechanics of the procedure.

Pre-Lab Discussion

Thoroughly discuss the results of the Exploration. Be sure to work through a sample calculation of relative viscosity using the equation given in the Exploration. Emphasize that the viscosity values are relative values and not absolute values. If students are uncertain about how to proceed, you may provide the following leading questions to help them as they make their plans.

Sample Data and Analysis

Using the data from the Exploration and the viscosity of water given in the Investigation, the relative viscosity of SAE-10 oil is about 16.7 cp at about 5°C, and the relative viscosity of SAE-20 oil is about 23.5 cp. At room temperature, the oils with relative viscosities nearest to 10 cp are SAE-20 (8.1) and SAE-30 (12).

Composition	Time (s)	Mass (g)	Volume (mL)	Temp. (°C)
1:1 SAE-10 and 20	26.2	8.8	10.0	5.0
1:1 SAE-20 and 30	17.8	9.0	10.0	20.0
H_2O	1.6	10.0	10.0	20.0 and 4.0

Density
for 10-20 mixture:
$$\frac{8.8 \text{ g}}{10.0 \text{ mL}} = 0.88 \text{ g/mL}$$

for 20-30 mixture:
$$\frac{9.0 \text{ g}}{10.0 \text{ mL}} = 0.90 \text{ g/mL}$$

for H_2O:
$$\frac{10.0 \text{ g}}{10.0 \text{ mL}} = 1.00 \text{ g/mL}$$

Relative Viscosity
for 10-20 mixture:
$$\frac{0.88 \text{ g/mL} \times 26.2 \text{ s} \times 1.52 \text{ cp}}{1.0 \text{ g/mL} \times 1.6 \text{ s}} = 22 \text{ cp*}$$

* Students who chose a mixture with slightly more than half SAE-10 will have results closer to 20 cp.

for 20-30 mixture:
$$\frac{0.90 \text{ g/mL} \times 17.8 \text{ s} \times 0.98 \text{ cp}}{1.0 \text{ g/mL} \times 1.6 \text{ s}} = 9.8 \text{ cp}$$

- At what temperature was data measured in the Exploration?
- For what temperature were the relative viscosities calculated in the Exploration?
- At what temperature are the specifications needed in this Investigation?
- How close are the specifications needed to the oils actually tested in the Exploration?
- Using the graphs from the Exploration, what properties can you predict given the desired relative viscosity?
- Which oils from the Exploration are the best candidates for combining to achieve the desired specifications?
- Which is easiest to make and to predict characteristics of: a mixture of two oils, of three oils, or of four oils?
- For each set of oils, what are several ratios that are likely to give a mixture with properties similar to those specified?
- What will you do to make sure that the mixture of oils is truly homogeneous?

Be certain that students have carefully thought out which oils to test. They are unlikely to stumble on the proper solution through trial and error in the time allotted for the lab. Check that all necessary tests are accounted for in the procedure. Students should measure mass, volume, temperature, and flow time for all of their mixtures and for water. They should be calculating density and relative viscosity for each oil mixture.

Proposed Procedure

Compare the specifications to the data from the Exploration. The relative viscosity values desired fall between those of SAE-20 and SAE-30 oil for one case and SAE-10 and SAE-20 on the other. Combine these oils in several ratios, including 1:1. These oils should be thoroughly mixed because they are viscous enough that they do not immediately form a homogeneous mixture on being combined. For one mixture, measure the mass of an empty beaker, and then pour the combined oil mixture into a graduated cylinder. Measure the volume to the nearest 0.1 mL. Pour the oil mixture from the graduated cylinder into the empty beaker, and measure the total mass. Calculate density for the oil mixture. Repeat for the other mixture. Use an ice-water bath to cool the oil mixtures that need a viscosity of 20 cp at 5°C. Measure and record the temperature to the nearest 0.1°C. Using a stopwatch or clock with second hand, measure the flow time in a pipet as in the Exploration, controlling the flow by covering and uncovering a pinhole in the top of a pipet bulb. Repeat the test for the oil mixture that needs a viscosity of 10 cp at 21°C. No ice-water bath should be necessary. Also repeat the test for water.

Post-Lab

Disposal

Set out eight disposal containers, one for each of the six types of pure oil labeled SAE-10 through SAE-60, one for multigrade oil mixtures, and one for the viscosimeter pipets. Students in successive periods can reuse the labeled beakers, test tubes, and pipets. Any remaining oil of known SAE rating can be reused class after class and year after year. The mixtures should be disposed of according to local regulations for the disposal of oil. Do not bury it or pour it down the drain. If the oil is cleaned up with paper towels, they can be disposed of only in a dry landfill designated for hazardous waste.

Additional Notes

Chemical Reactions and Solid Fuel

EXPLORATION 7-1
Technique Builder

Objectives

Students will
- use appropriate lab safety procedures.
- use a laboratory balance to measure mass.
- use a graduated cylinder to measure volume.
- demonstrate proficiency in vacuum or gravity filtration.
- observe several chemical reactions and physical changes.
- identify evidence that a chemical change has occurred.
- recognize several different reaction types.
- write balanced chemical equations for reactions observed.
- prepare a gel and determine whether it is homogeneous or heterogeneous.
- calculate the mass of a fuel before and after burning.
- evaluate lab-produced gelled or solid fuel by measuring how well it heats water compared to a commercially available solid fuel.

Planning

Recommended Time

2 lab periods (ethanol fuel can be analyzed and tested on one day, and isopropanol analyzed and tested on the following day)

Materials

(for each lab group)
- Chalk, 25 g
- Commercially available solid fuel, such as Sterno
- Distilled water
- Ethanol, CH_3CH_2OH, 30 mL
- Isopropanol, $CH_3CHOHCH_3$, 30 mL
- Vinegar (acetic acid solution), 100 mL
- 250 mL beakers, 6
- 100 mL graduated cylinder
- Balance
- Crucible tongs
- Evaporating dishes, 2
- Glass stirring rod
- Mortar and pestle
- Ring stands with ring clamps, 3
- Wire gauze with ceramic center, 3
Teacher use only
- Safety matches

continued on next page

Required Precautions

- Goggles and a lab apron must be worn at all times.
- Read all safety cautions, and discuss them with your students.
- The isopropanol and ethanol are extremely flammable. They should be kept within an operating fume hood. The bottles should be closed when not in use. Place only 300 mL at a time in the reagent bottle, with a lid or stopper nearby. Students should replace the lid when they are finished. Closely supervise the dispensing of alcohol and the return of excess to the waste alcohol containers.
- Do not dispense any isopropanol or ethanol while there are burners, flames, hot plates, or other heat sources in use.
- Before lighting the fuels to test them, be sure all excess ethanol and isopropanol have been placed in the fume hood with the window closed and the fan on.

continued on next page

Required Precautions

(continued)

- Clean up any spills with a damp cloth, keeping the cloth in the hood.
- If a hot plate or the LEAP System with thermistor probe are used, the precautions listed in the teacher's notes on page T37 must be followed to avoid electric shock.

Gravity filtration option
- Glass funnel
- Filter paper

Vacuum filtration option
- Aspirator for spigot
- Büchner funnel (either ceramic or plastic)
- Filter paper
- One-hole rubber stopper or sleeve
- Vacuum flask, plastic (sidearm flask), and tubing

Bunsen burner option
- Bunsen burner
- Gas tubing
- Striker

Hot plate option
- Hot plate

LEAP System option
- Thermistor probes, 3

Thermometer option
- Thermometers, nonmercury, 3

Solution/Material Preparation

1. Be certain that the ethanol and isopropanol are kept in stock bottles with lids under the fume hood. Do not dispense any alcohol until all students have finished completely the heating of the chalk-vinegar filtrate and all burners or hot plates have been turned off and have cooled down. Make only 300 mL of ethanol or isopropanol available at a time. Students should replace the lid as soon as they have taken what they need. Once the students have all made their gelled fuels, do not proceed until all students have drained any excess alcohol into the appropriate disposal containers, and the containers have been placed in the fume hood. If the alcohol is poured only in the hood, the likelihood of an unplanned combustion are greatly reduced. To maximize safety, keep hot plates and

burners set up in the half of the room opposite the hood.

2. Only one ring stand per lab group is necessary if the preparations and testing of the different fuels are performed on separate days, and the commercial fuel is burned and tested as a demonstration at the beginning or end of the class.

3. The safety matches indicated for teacher use only are for lighting the gelled fuels. Be certain to follow the required precautions before lighting the fuels. Only safety matches should be used in the lab. To maintain control, be certain that you, the teacher, light the students' samples when they are ready to proceed. **Do not permit students to light the fuel themselves.**

4. If you choose the vacuum filtration option, do not use a glass sidearm flask. Polypropylene filter flasks with sidearms are available from many suppliers.

5. Be sure to use anti-glare chalk. This type of chalk seems to be the only kind that produces a saturated solution of calcium acetate.

6. Use vinegar purchased through a chemical supply company or a grocery store instead of diluting glacial acetic acid.

Student Orientation

Techniques to Demonstrate

Make sure students appreciate the importance of waiting to work with the alcohols until everyone has finished the heating step and their equipment has cooled down. Similarly, all students should wait for everyone to completely finish making their gelled fuel with alcohol before they ask for you to light the fuel for testing. The isopropanol will make less gel than the ethanol. Be certain all excess alcohol is poured off into disposal containers that have been placed in the hood before matches or any other heat source is used in the lab.

Review with students the proper technique for grinding material in a mortar and pestle. Be sure students check to make sure that the filtrate is clear. If it is not, a re-filtering step may be necessary.

Pre-Lab Discussion

Students may need to be reminded of the distinguishing features of homogeneous and heterogeneous mixtures. The gel is essentially a heterogeneous mixture and not a new chemical substance.

Discuss with students the principles of experimental design while setting up the comparison study with commercial solid fuel. For example, the ring stand arrangements should be identical, with similar volumes of water, similarly sized beakers, etc.

Sample Data

Measurement	Ethanol fuel	Isopropanol fuel
Mass of empty evap. dish (g)	47.20	48.15
Mass of gel + dish (g)	77.92	68.76
Mass of ash + dish (g)	47.93	48.74

Time (min)	Temp. (°C)— ethanol fuel	Temp. (°C)— isopropanol fuel	Temp. (°C)— comm'l solid fuel
0	21	21	21
1	23	22	22
2	30	26	24
3	32	31	26
4	40	40	28
5	44	43	29
6	48	45	30
7	52	48	31
8	54	51	33
9	56	53	34
10	63	59	36
11	66	56	38
12	65	56	40
13	63	56	42
14	63	55	43
15	63	55	44

Students should note some or all of the following observations in their data table. The addition of vinegar to the chalk caused bubbling. There was no evidence of a reaction when the resulting solution was heated. The gel appeared to be a heterogeneous mixture. Some excess isopropanol was left after the gel was formed. After burning, a white ashlike substance remained in the evaporating dish.

Post-Lab

Disposal

Three disposal containers should be set out before lab work begins. One is for excess ethanol, one is for excess isopropanol, and the last is for the ash (calcium acetate) that is left after the fuel burns. The filter paper and solid (which is mostly calcium carbonate) can be disposed in the trash, as can the ash (calcium acetate). Any calcium acetate solution remaining can be washed down the drain. Ethanol and isopropanol should be diluted with ten times their volume of water, and then washed down the drain.

Answers to

Analysis and Interpretation

1. The bubbles that formed when vinegar was added to chalk indicated that a gaseous product was being created, one of the signs of a chemical change.

2. $CaCO_3(s) + 2CH_3COOH(aq) \longrightarrow H_2CO_3(aq) + Ca(CH_3COO)_2(aq)$

3. $H_2CO_3(aq) \longrightarrow H_2O(l) + CO_2(g)$
The gas bubbles observed were carbon dioxide.

4. There was no evidence that heating the filtrate caused a reaction.

5. The combination produced neither heat, light, noise, a solid precipitate, nor a gaseous product. It is unlikely that a chemical change occurred.

6. The gel appears to be a heterogeneous mixture because, unless it is stirred, it does not have a similar form throughout.

7. ethanol: $2CH_3CH_2OH(l) + 7O_2(g) \longrightarrow 4CO_2(g) + 6H_2O(l)$
isopropanol: $CH_3CHOHCH_3(l) + 5O_2(g) \longrightarrow 3CO_2(g) + 4H_2O(l)$

8. No, the ash left over is not a true combustion product. It is the calcium acetate that is left over after everything else has burned.

9. $\dfrac{0.73 \text{ g Ca(CH}_3\text{COO)}_2}{30.72 \text{ g ethanol fuel}} = 2.4\%$

$\dfrac{0.59 \text{ g Ca(CH}_3\text{COO)}_2}{20.61 \text{ g isopropanol fuel}} = 2.9\%$

10. $30.72 \text{ g gel} - 0.73 \text{ g Ca(CH}_3\text{COO)}_2 =$
$\qquad\qquad\qquad\qquad 29.99 \text{ g ethanol}$

$29.99 \text{ g ethanol} \times \dfrac{1 \text{ mol}}{46.08 \text{ g}} =$
$\qquad\qquad\qquad 0.6508 \text{ mol ethanol}$

$20.61 \text{ g gel} - 0.59 \text{ g Ca(CH}_3\text{COO)}_2 =$
$\qquad\qquad\qquad 20.02 \text{ g isopropanol}$

$20.02 \text{ g isopropanol} \times \dfrac{1 \text{ mol}}{60.11 \text{ g}} =$
$\qquad\qquad\qquad 0.3331 \text{ mol isopropanol}$

11. $0.6508 \text{ mol} \times \dfrac{-1367 \text{ kJ}}{1 \text{ mol}} =$
$\qquad\qquad\qquad -889.6 \text{ kJ for ethanol}$

$0.3331 \text{ mol} \times \dfrac{-2006 \text{ kJ}}{1 \text{ mol}} =$
$\qquad\qquad\qquad -668.2 \text{ kJ for isopropanol}$

More energy should have been provided by ethanol.

12.

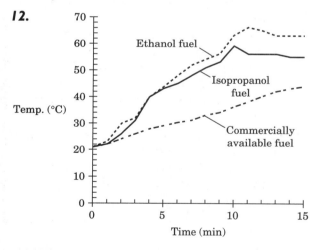

Temp. (°C) vs. Time (min)

Ethanol fuel
Isopropanol fuel
Commercially available fuel

Conclusions

13. Ethanol was the most effective fuel for heating water. The commercially available fuel was the least effective for heating water.

14. The fuel that ignites easily, quickly, and with a lot of heat could also be more

dangerous because it could ignite accidentally. Also, it might burn out too quickly to be useful in heating objects.

15. Neither hydrocarbons nor nitrogen oxide compounds are products of the main reactions involved in this process.

Extensions

1. Δt ethanol for 75 mL H_2O = 45°C

$79°C \times \dfrac{30.72 \text{ g ethanol fuel}}{45°C} \times$

$\dfrac{250 \text{ mL}}{75 \text{ mL}} = 180 \text{ g ethanol fuel}$

Δt isopropanol for 75 mL H_2O = 38°C

$79°C \times \dfrac{20.61 \text{ g isopropanol fuel}}{38°C} \times$

$\dfrac{250 \text{ mL}}{75 \text{ mL}} = 140 \text{ g isopropanol fuel}$

2. For the ethanol fuel, about six times as much fuel would be required. The chalk and vinegar were enough to make two batches, so three times as much of each of them would be needed.

$75.0 \text{ g chalk} \times \dfrac{\$6.25}{1000 \text{ g}} = \qquad \$0.469$

$300 \text{ mL vinegar} \times \dfrac{\$0.86}{1000 \text{ mL}} = \quad \$0.2580$

$180.0 \text{ mL ethanol} \times \dfrac{\$3.27}{1000 \text{ mL}} = \underline{\$0.589}$

$\qquad\qquad\qquad \text{unit price} = \$1.32$

For the isopropanol fuel, about four times as much fuel would be required, so twice as much chalk and vinegar would be needed.

$50.0 \text{ g chalk} \times \dfrac{\$6.25}{1000 \text{ g}} = \qquad \$0.312$

$200.0 \text{ mL vinegar} \times \dfrac{\$0.86}{1000 \text{ mL}} = \$0.1720$

$120.0 \text{ mL isopropanol} \times \dfrac{\$4.62}{1000 \text{ mL}} =$

$\qquad\qquad\qquad\qquad \underline{\$0.554}$
$\qquad\qquad \text{unit price} = \$1.038$

Specific Heat Capacity

Objectives

Students will
- use appropriate lab safety procedures.
- build a simple calorimeter.
- calibrate their calorimeter with accurately measured amounts of hot and cold water.
- measure temperature using a thermometer or a LEAP System thermistor probe.
- measure volume using a graduated cylinder.
- measure mass using a laboratory balance.
- relate measurements of temperature to changes in heat content.
- heat a metal sample with a hot-water bath.
- use a calorimeter to determine the specific heat capacity of metals.

Planning

Recommended Time

1 lab period

Materials

(for each lab group)
- Aluminum metal, 75 g
- Copper metal, 75 g
- Iron metal, 75 g
- Distilled water
- 400 mL beakers, 2
- 100 mL graduated cylinder
- Balance
- Beaker tongs
- Boiling chips
- Glass stirring rod
- Plastic foam cups for calorimeter, 2
- Scissors
- Ring stand
- Test-tube holder
- Test tubes, large or medium, 3

Bunsen burner option
- Bunsen burner and related equipment
- Ring clamp
- Wire gauze with ceramic center

Hot plate option
- Hot plate

LEAP System option
- LEAP System
- LEAP thermistor probe

Thermometer option
- Thermometer, nonmercury

Required Precautions

- Goggles and a lab apron must be worn at all times to provide protection for the eyes and clothing.
- Tie back long hair and loose clothing.
- Read all safety cautions, and discuss them with your students.
- Remind students to use beaker tongs when handling the beaker containing hot water because it can burn or scald.
- If the LEAP System with thermistor probe or a hot plate are used, the precautions listed in the teacher's notes on page T37 must be followed to avoid electric shock.

Solution/Material Preparation

1. Smaller amounts of metals can be used with this lab if less accuracy is acceptable. Rather than reagent-grade metals, try to locate inexpensive sources, such as iron nails, discarded copper wire, and aluminum pellets. Be sure to eliminate any sharp edges.
2. Once a calorimeter has been made, it can be kept and reused by successive classes and for other experiments.

3. Tea infusers from kitchen supply stores can be used instead of test tubes to heat the metal samples. They allow quicker and more complete heating of the metal, and the little water that is left on them during the transfer does not appreciably alter the results.

Student Orientation

At this point in the year, students should be familiar with all of the equipment necessary for this lab. Be sure to emphasize that thermometers are too fragile to be used as stirring rods.

Make certain students understand that the more air-tight their calorimeter is, the less heat energy will escape. Demonstrate how to quickly transfer the hot metal sample from the test tube to the calorimeter without splashing water out of the calorimeter.

Better results can be obtained by repeating the calibration steps to determine an aver-

age value for the specific heat of the calorimeter, C', but this may require two lab periods.

Many students have difficulty relating the gain in heat of the water in the calorimeter to the loss in heat of the metal. Refer back to the discussion of the conservation of energy and the flow of heat energy as needed to help them relate this specific idea to a larger context.

Post-Lab

Set out three disposal containers, one for each kind of metal. All water can be poured down the drain. The metals should be saved and reused. Be certain that they are well dried before storing them.

Answers to

1. The assumption that energy released by one part of the system is absorbed by another part of the system is based on the law of conservation of energy, which states that energy can be transferred from one object to another or converted from one form to another but cannot be created or destroyed.

2. $50.0 \text{ mL cool } H_2O \times \dfrac{1.00 \text{ g}}{1 \text{ mL}} =$

$$50.0 \text{ g cool } H_2O$$

$50.0 \text{ mL hot } H_2O \times \dfrac{0.97 \text{ g}}{1 \text{ mL}} =$

$$48.5 \text{ g hot } H_2O$$

3. $\Delta t_{cool\ H_2O} = 49.5°C - 25.0°C = 24.5°C$
$\Delta t_{hot\ H_2O} = 75.0°C - 49.5°C = 25.5°C$

4. change in heat energy for cool H_2O =
$(4.180 \text{ J/g} \cdot °C)(50.0 \text{ g})(24.5°C) =$

$$5.12 \text{ kJ}$$

change in heat energy for hot H_2O =
$(4.180 \text{ J/g} \cdot °C)(48.5 \text{ g})(25.5°C) =$

$$5.17 \text{ kJ}$$

5. The values should be slightly different because some heat energy will be lost to the surroundings. That is why not all of the heat energy released by the hot water was absorbed by the cold water. The heat lost to the surroundings was 0.05 kJ, or 50 J.

Sample Data

Calibration of Calorimeter

Measurement	Cool H_2O	Hot H_2O	Calorimeter
Volume (mL)	50.0	50.0	X
Init. temp. (°C)	25.0	75.0	25.0
Final temp. (°C)	49.5	49.5	49.5

Testing of Metals

Measurement	Al	Cu	Fe
Test-tube mass (g)	28.0	29.5	28.8
Metal + test-tube mass (g)	63.2	116.7	98.0
Metal init. temp. (°C)	99.0	99.0	99.0
H_2O volume (mL)	75.0	75.0	75.0
H_2O init. temp. (°C)	21.0	21.0	21.0
Final temp. (°C)	28.2	28.2	28.1

6. $C' = \dfrac{0.05 \text{ kJ}}{24.5°C} = 2 \text{ J/°C}$

7. The lower the value of C', the less heat energy will escape into the surroundings, and the more accurate any calculations of heat changes are likely to be.

8. 63.2 g − 28.0 g = 35.2 g Al
116.7 g − 29.5 g = 87.2 g Cu
98.0 g − 28.8 g = 69.2 g Fe

9. Aluminum test
28.2°C − 21.0°C = 7.2°C = Δt_{H_2O}
99.0°C − 28.2°C = 70.8°C = Δt_{Al}
Copper test
28.2°C − 21.0°C = 7.2°C = Δt_{H_2O}
99.0°C − 28.2°C = 70.8°C = Δt_{Cu}
Iron test
28.1°C − 21.0°C = 7.1°C = Δt_{H_2O}
99.0°C − 28.1°C = 70.9°C = Δt_{Fe}

Conclusions

10. heat energy released by metal = heat energy absorbed by H_2O + heat energy released to surroundings

11. heat energy absorbed by H_2O = $m_{H_2O} \times c_p \times \Delta t_{H_2O}$
heat energy released to surroundings = $C' \times \Delta t_{H_2O}$
(Note that the change in temperature for the water is assumed to be the change in temperature for the calorimeter.)

12. Aluminum test
heat energy absorbed by H_2O =
75.0 g × 4.180 J/g°C × 7.2°C = 2.3 kJ
heat energy released to surroundings:
2 J/°C × 7.2°C = 14 J = 0.014 kJ
Copper test
heat energy absorbed by H_2O =
75.0 g × 4.180 J/g°C × 7.2°C = 2.3 kJ
heat energy released to surroundings:
2 J/°C × 7.2°C = 14 J = 0.014 kJ
Iron test
heat energy absorbed by H_2O =
75.0 g × 4.180 J/g°C × 7.1°C = 2.2 kJ
heat energy released to surroundings:
2 J/°C × 7.1°C = 14 J = 0.014 kJ

13. Aluminum test
energy released by metal = 2.3 kJ + 0.014 kJ = 2.3 kJ, to correct significant figures
2.3 kJ = $c_{p,\,Al}$(35.2 g Al)(70.8°C)
$c_{p,\,Al}$ = 0.92 J/g°C
Copper test
energy released by metal = 2.3 kJ + 0.014 kJ = 2.3 kJ, to correct significant figures
2.3 kJ = $c_{p,\,Cu}$(87.2 g Cu)(70.8°C)
$c_{p,\,Cu}$ = 0.37 J/g°C
Iron test
energy released by metal = 2.2 kJ + 0.014 kJ = 2.2 kJ, to correct significant figures
2.2 kJ = $c_{p,\,Fe}$(69.2 g Fe)(70.9°C)
$c_{p,\,Fe}$ = 0.45 J/g°C

14. $\dfrac{0.92 - 0.90}{0.90} \times 100 = 2\%$ for Al

$\dfrac{0.37 - 0.38}{0.38} \times 100 = 3\%$ for Cu

$\dfrac{0.45 - 0.44}{0.44} \times 100 = 2\%$ for Fe

15. Metals have low specific heat capacities compared with water. The temperature of the metals will change more than the water temperature.

16. Copper is the metal that heats up the quickest because it has the lowest specific heat capacity.

Extensions

1. Students' answers will vary depending on class answers. Make sure calculations of averages are accurate. Usually, the class average is closer than many of the individual measurements.

2. Students' suggestions for improving the calorimeter's insulation will vary. Be sure answers are safe and include carefully planned procedures.

3. To prevent burns, cookware handles should be made of substances with high specific heat capacities, which can absorb considerable energy with small changes in temperature.

Specific Heat Capacity

Objectives

Students will
- use appropriate lab safety procedures.
- build and calibrate their calorimeter with accurately measured amounts of hot and cold water.
- measure temperature using a thermometer or a LEAP System thermistor probe.
- measure volume using a graduated cylinder.
- measure mass using a laboratory balance.
- relate measurements of temperature to changes in heat content.
- heat a metal sample with a hot-water bath.
- use a calorimeter to determine the specific heat capacity of metals.
- identify a metal based on its specific heat capacity.
- identify another physical property that can be used to identify a metal.
- design and implement their own procedure.

Planning

Recommended Time
1 lab period

Prerequisite
Exploration 9-1

Materials
(for each lab group)
- Metal sample, about 25 g
- 400 mL beakers, 2
- 100 mL graduated cylinder
- Balance
- Beaker tongs
- Chemical handbook
- Boiling chips
- Glass stirring rod
- Magnets
- Plastic foam cups for calorimeter, 2
- Scissors or tools to trim cups
- Ring stand
- Test-tube holder
- Test tube, large or medium

Bunsen burner option
- Bunsen burner and related equipment
- Ring clamp
- Wire gauze with ceramic center

Hot plate option
- Hot plate

LEAP System option
- LEAP System
- LEAP thermistor probe

Thermometer option
- Thermometer, nonmercury

Required Precautions

- Goggles and a lab apron must be worn at all times to provide protection for the eyes and clothing.
- Tie back long hair and loose clothing.
- Read all safety cautions, and discuss them with your students.
- Remind students to use beaker tongs when handling the beaker containing hot water because it can burn or scald.
- If the LEAP System with thermistor probe or a hot plate are used, the precautions listed in the teacher's notes on page T37 must be followed to avoid electric shock.

Estimated cost of materials: $97 500 (less if students do not need to rebuild and recalibrate a calorimeter)

Solution/Material Preparation

1. The metal sample can be any of those mentioned in the letter, but the best choices are iron or aluminum because students will then have a chance to compare the specific heat capacity value of the unknown to their data from the Exploration. Do not provide these metals in

the same form as in the Exploration. It is even better if different forms of the metal are provided to the lab groups to cut down considerably on copying other's results instead of doing the lab.

2. Students can save time by reusing the calorimeters made in the Exploration especially since there will be no need for the calibration step if they did an accurate job in the Exploration.

3. Do not provide standard samples of the other metals. Instead, make students rely on values from a chemical handbook or other reference work. The *CRC Handbook of Chemistry and Physics* or *Lange's Handbook of Chemistry* will have the necessary information. Alternatively, you can give students a fact sheet listing the properties of several metals, such as specific heat capacity, density, color, magnetic effects, electronegativity, electron affinity, ionization energy, etc.

Student Orientation

Be sure students remember the necessary steps from the Exploration. Emphasize the need to be certain that the metal has been in the hot-water bath long enough to be at the same temperature as the water. Remind students that they should never use thermometers to stir anything.

Begin by discussing the results and procedure used in the Exploration, especially any errors in lab technique that occurred. Be certain students understand how to apply the specific heat capacity equations from the Exploration to their data.

Be certain that students indicate exactly which property, besides specific heat capacity, they will measure for the sample. Density is the most accurate method for identifying the sample. The most common error is not drying the metal before measuring its mass. The iron samples can be identified by the property of magnetism.

Proposed Procedure

Measure the mass of the piece of metal. Place exactly 25.0 mL of water in a graduated cylinder. Add the metal to the graduated cylinder, and record the final volume to the nearest 0.1 mL. Build and calibrate a calorimeter, as described in the Exploration, or use one that has already been made and calibrated. Measure exactly 50.0 mL of water in the graduated cylinder. Pour it into the calorimeter, monitoring the temperature until it is constant. Record this temperature to the nearest 0.1°C as the initial water temperature. Prepare a hot-water bath with a 400 mL beaker full of water and a hot plate or a Bunsen burner with ring stand and ring. Place the metal in a large test tube, attaching the test tube to a ring stand with a test-tube clamp. Lower the test tube into the water until the metal is below the surface of the water. Wait about 10 min, and then record the water temperature to the nearest 0.1°C as the initial metal temperature. Quickly transfer the metal to the calorimeter and stir the contents for 30 s. Record the highest temperature attained as the final temperature.

Post-Lab

Set out a disposal container for the metal samples. The metal samples can be reused if they are dried every time this Investigation is performed.

Sample Data and Analysis

The sample data is for aluminum.

Metal mass (g)	22.5
Metal volume (mL)	$33.3 - 25.0 = 8.3$
Metal init. temp. (°C)	100.0
H_2O volume (mL)	50.0
H_2O init. temp. (°C)	21.0
Final temp. (°C)	28.0

$$\text{density of metal} = \frac{22.5 \text{ g}}{8.3 \text{ mL}} = 2.7 \text{ g/mL}$$

$\Delta t_{H_2O} = 7.0°C \qquad \Delta t_{metal} = 72.0°C$

heat energy absorbed by $H_2O =$
   $(50.0 \text{ g})(4.180 \text{ J/g} \cdot °C)(7.0°C) = 1.5 \text{ kJ}$

heat energy released by metal $= -1.5 \text{ kJ} =$
   $(22.5 \text{ g})(c_{p,metal})(72.0°C)$

$c_{p,metal} = 0.93 \text{ J/g} \cdot °C$

These values most closely match aluminum, which implicates the Spleen Activist Group.

Constructing a Heating/ Cooling Curve

9-2
Technique Builder

Objectives

Students will
- use appropriate lab safety procedures.
- observe and record temperature changes as a pure substance melts and freezes.
- measure time using a stopwatch or a clock with a second hand.
- measure temperature using a thermometer or a LEAP System with thermistor probe.
- graph and analyze time and temperature data.
- determine the melting/freezing point of a pure substance.
- identify the relationship between temperature and a phase change.
- infer the relationship between heat energy and phase changes based on data.
- analyze the interrelationships among energy, entropy, and temperature.

Planning

Recommended Time

1 lab period

Materials

(for each lab group)
- Ice
- $Na_2S_2O_3 \cdot 5H_2O$, 15 g
- 600 mL beakers, 3
- Balance, centigram
- Beaker tongs
- Chemical reference books
- Forceps
- Graph paper
- Hot mitt
- Plastic washtub
- Ring stand
- Ring clamps, 3 (One must be large enough for a 600 mL beaker to fit inside it.)
- Ruler
- Stopwatch or clock with a second hand
- Test-tube clamp
- Test tube, Pyrex medium
- Thermometer clamp
- Wire gauze with ceramic center, 2
- Wire, 25 cm

Bunsen burner option
- Bunsen burner
- Gas tubing
- Striker

Required Precautions

- Goggles and a lab apron must be worn at all times.
- Read all safety cautions, and discuss them with your students.
- Make sure the iron rings are large enough to hold a 600 mL beaker.
- If a hot plate or the LEAP System with thermistor probe is used, the precautions listed in the teacher's notes on page T37 must be followed to avoid electric shock.

Hot plate option
- Hot plate

LEAP System option
- LEAP System
- Thermistor probes, 2

Thermometer option
- Thermometers, nonmercury, 2

Solution/Material Preparation

1. This lab will go more quickly if several wire stirrers are prepared in advance. Cut 25 cm lengths of wire. (Any gauge will do provided it is easily bent.) Make a loop at one end that has a 1 cm diameter. Bend the wire where it attaches to the loop so that the loop is perpendicular to the rest of the wire. Bend the top point of the wire over in a small loop to use as a handle.

2. Fill test tubes with about 15.0 g of $Na_2S_2O_3 \cdot 5H_2O$. The solid can be reused several times. Measure out one 15 g sample, completely transfer it to a test tube, and fill the remaining test tubes to the same level.

3. Be sure to set out a wide-mouthed bottle containing several small crystals of $Na_2S_2O_3 \cdot 5H_2O$ for students to use as seed crystals.

4. Other test-tube clamps may be used in place of the three-fingered one.

Student Orientation

Techniques to Demonstrate

Students find it particularly awkward to manipulate the setups, so a quick run-through of the steps may prove beneficial: assembling and operating the hot-water bath, positioning the thermometer in the solid, raising and lowering and exchanging the beakers, and when to begin and end timing.

The use of the LEAP thermistor is somewhat complicated for the second melting test because of the difficulty in reaching a plateau when the solid melts. You may have to demonstrate how to position the burner to keep the temperature of the water bath constant at around 60°C.

Pre-Lab Discussion

Remind students that thermometer bulbs must not rest on the bottom of test tubes or beakers and that the thermometers must never be used to stir anything. The most accurate temperature readings are made when the thermometer is vertical and the line of sight is horizontal, not angled.

Point out the purpose of the initial quick melt. Emphasize that only one temperature is taken during the quick melt. This provides a rough idea of where the melting and freezing point will occur. In the succeeding tests, more attention should be paid to the temperature readings near this quick-melt temperature, and repetitive readings should be expected.

Explain that the temperature of a pure substance remains constant during a phase change. Review what happens at the particle level during melting and freezing.

NOTE: The $Na_2S_2O_3 \cdot 5H_2O$ will undercool unless a seed crystal is added a few degrees above its freezing temperature. The temperature may still dip one or two degrees below the freezing temperature, but it will rise when the liquid is stirred.

Post-Lab

Disposal

Remelt the sodium thiosulfate pentahydrate in a water bath, pour all of the liquid in a wide-mouth reagent jar, cool to room temperature, cover, and label for reuse. It will be necessary to pulverize the crystals into smaller chunks before reusing them.

Answers to

Analysis and Interpretation

1.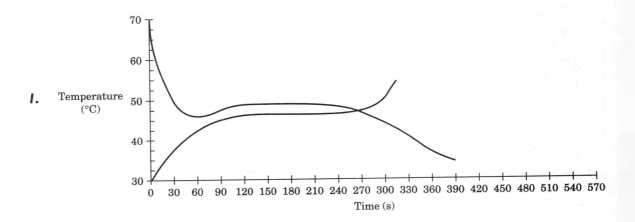

2. The cooling curve begins at an initial high temperature, slopes rapidly downward to a plateau at 48.2°C, and then continues to slope downward. The heating curve begins at an initially low temperature, slopes rapidly upward to a plateau, and then continues to slope upward. The two curves overlap in the plateau area.

Sample Data and Analysis

Cooling Data			Warming Data		
Time (s)	Temp. (°C)	Observ.	Time (s)	Temp. (°C)	Observ.
0	69.0		0	29.0	
15	61.0		15	31.5	
30	51.0	Seed crystal added	30	36.0	
45	47.0	Crystallization	45	39.5	
60	46.0		60	42.0	
75	48.2		75	43.5	
90	48.2		90	44.5	
105	48.2		105	45.5	
120	48.2		120	46.0	Melting starts
135	48.2		135	46.5	
150	48.2		150	47.0	
165	48.2		165	47.0	
180	48.2		180	47.2	
195	48.2		195	47.4	
210	48.2		210	47.5	
225	48.2		225	47.5	
240	48.0	Total solid	240	47.5	
255	47.5		255	48.0	
270	47.0		270	55.0	All solid melted
285	45.7		285	55.0	
300	44.7		300	58.0	
315	43.5		315	60.0	
330	41.5				
345	39.0				
360	36.2				
375	34.0				
390	32.0				

3. The melting point is 48.2°C, while the freezing point is 47.5°C. They are close but should be the same. The difference in names reflects the phase change that is taking place and tells from which direction the phase change is being approached.

4. The quick melt provides a rough measure of the melting temperature. It tells the experimenter to pay close attention to time and temperature measurements near this point because the phase change plateau should occur.

5. from Merck Index: m.p. = 48°C when rapidly heated

$$\frac{48.2 - 48}{48} \times 100 = 1.67\%$$

Conclusions

6. As the liquid cools, the kinetic energy and entropy decrease for the $Na_2S_2O_3 \cdot 5H_2O$ particles but increase for the water bath.

7. The temperature remained relatively constant, but the entropy decreased while the liquid was freezing into a solid.

8. The larger the sample size, the longer the sample remains at the plateau temperature.

9. No, melting point is an intensive property, so it depends on the nature of the substance being melted, not on the quantity.

10. The substance is impure because the melting point is different; Δt is about 3°C.

Extensions

1. The energy and entropy decrease during cooling but increase during heating. The downward slope seems to indicate a large transfer of energy to the surroundings, so molecules are moving less rapidly; the upward slope indicates a large transfer from the surroundings, so molecular motion increases. The phase changes during plateaus represent times at which there are no changes in temperature and only small changes in molecular motion, even though substantial changes occur in energy and entropy.

2. The thermostat should turn on the freezing-rain defroster when the temperature is only a few degrees above freezing (0°C). The defroster should be turned off when the temperature is warm enough that the freezing rain should melt immediately, at approximately 20°C.

3. Crystallization should eventually take place when the temperature is below the freezing point. However, this process will not be as rapid as when the cooling water bath and seed crystal are used.

4. A water bath warms the test tube evenly. Distilled water is not necessary because the water does not mix with the chemicals.

5.
   __a__ 1. ice melting at 0°C
   __b__ 2. water freezing at 0°C
   __d__ 3. a mixture of ice and water whose relative amounts remain unchanged
   __c__ 4. a particle escaping from a solid and becoming a vapor particle
   __e__ 5. solids, like camphor and naphthalene, evaporating
   __f__ 6. snow melting
   __c__ 7. snow forming
   __c__ 8. dry ice subliming
   __f__ 9. dry ice forming

Constructing a Cooling Curve—Melting Oils for Soap Production

INVESTIGATION 9-2

Problem Solving

Objectives

Students will
- use appropriate lab safety procedures.
- measure the volume of a sample of water using a graduated cylinder.
- calculate the mass of water using density.
- measure the mass of fat samples with a laboratory balance.
- observe temperature changes when fats are melted in a calorimeter using a thermometer or LEAP System thermistor probe.
- determine the melting points of several fats.
- collect data on the relationship between heating time and temperature for three fats.
- create a heating/cooling curve from the data.
- apply the specific heat capacity equation to relate temperature changes to enthalpy changes.
- rank the fats in order of their heats of fusion.
- design and implement their own procedure.

Planning

Recommended Time
1–2 lab periods. Some fats can be tested on one day, others on the next day.

Prerequisites
Exploration 9-2

Materials
(for each lab group)
- Tallow, 10.0 g
- Vegetable shortening, 10.0 g
- Coconut oil, 10.0 g
- Distilled water
- Ice
- 400 mL beakers, 2
- 100 mL graduated cylinder
- Balance
- Beaker tongs
- Glass stirring rod
- Stopwatch or clock with second hand
- Test-tube holder
- Test-tube rack
- Test tubes, large, 3

continued on the next page

Required Precautions

- Goggles and a lab apron must be worn at all times to provide protection for the eyes and clothing.
- Tie back long hair and loose clothing.
- Read all safety cautions, and discuss them with your students.
- Remind students to use beaker tongs when handling the beaker containing hot water because it can burn or scald.
- Remind students that the fats should be heated only in a water bath, not directly over a flame.
- Do not use any of these fat samples from the lab to prepare food.
- If the LEAP System with thermistor probe or a hot plate is used, the precautions listed in the teacher's notes on page T37 must be followed to avoid electric shock.

Bunsen burner option
- Bunsen burner and related equipment
- Ring clamp
- Ring stand
- Wire gauze with ceramic center

Hot plate option
- Hot plate

LEAP System option
- LEAP System
- LEAP thermistor probe

Thermometer option
- Thermometer, nonmercury

Estimated cost of materials:
$120 000–$130 000

Solution/Material Preparation

1. Coconut oil and tallow can be obtained from many chemical supply companies. If tallow is unavailable, lard may be substituted.
2. The hot water is required to be at a temperature of only 50°C. If the hot tap water is that hot, use it rather than heating cool water with a Bunsen burner or hot plate.

Student Orientation

Techniques to Demonstrate

Demonstrate melting a small sample of fat by placing it in a test tube within a hot-water bath. The melting and solidifying of the fat can be controlled just as was done with the thiosulfate crystals in Exploration 9-2. However, no seed crystal is required for the fats.

Pre-Lab Discussion

Thoroughly discuss Exploration 9-2, especially any errors in lab technique that occurred. Encourage students to sketch a graph of their data before they leave the lab. That way they will be able to identify whether their results are conclusive, or whether additional trials will be necessary. It is not necessary to produce a heating curve for this investigation. These fats tend to have a range of about 5°C across which melting occurs, rather than a single, identifiable temperature at which the phase-change occurs. As a result, the graphs have a different shape than those for pure substances. Remind students that as the fats reach the temperature of the water, the temperature will level off. The melting point should be a brief plateau that occurs earlier.

If students are uncertain about how to proceed, you may provide the following leading questions for them to consider as they make their plans.
- How do you know when to stop heating the fat samples?
- How do you know when to stop collecting data for the cooling curve?
- What is the relationship between the amount of heat energy measured for cooling down the melted liquid from its melting point to room temperature, and the amount of heat energy necessary to warm up a solid fat from room temperature to just beyond its melting point?

Evaluating the Pre-Lab Requirements

Be certain that students recognize the differences between this Investigation and the Exploration that preceded it. For example, there is no need to heat these fats to 85°C because they melt at much lower temperatures. Students' plans should indicate that they will heat the fats only until they are melted.

The proposed data tables should contain spaces for the temperature at which melting occurred, the initial temperature of the melted fat, the initial temperature of the cool water, and at least 20 temperature measurements taken at 15 s intervals. Students should indicate that they will use at least 10 g of each fat. Some may want to use more, but those proposing to use less should be persuaded to increase the amount.

Proposed Procedure

Measure and record the masses of three large test tubes. Add about 10.0 g of fat to each one. Measure and record the masses of the filled test tubes. Heat one sample of fat in its test-tube in a warm-water bath until it melts. Measure and record the temperature of the fat to the nearest 0.1°C after it melts. Fill a beaker with water whose temperature is 15°C or lower. Add about 150 mL of the cold water to a graduated cylinder. Pour the water into the open calorimeter cup, measuring and recording the temperature to the nearest 0.1°C. Rest the test tube of melted fat in the water, and measure the temperature every 15 s until it levels off. Observe the fat closely to determine the approximate melting/solidifying point. Repeat the process for the other fats.

Post Lab

Set out three different disposal containers, one for each kind of fat. If they are kept in sealed containers, these samples may be reused each time this Investigation is per-formed during a school year. Storing the samples from year to year may not be practi-cal, as they can grow rancid. To dispose of the samples when your students have fin-ished performing the Investigation for the year, place them in the trash.

Sample Data and Analysis

Time (s)	Tallow—Temp. (°C)	Veg. Short.—Temp. (°C)	Coconut Oil—Temp. (°C)	Time (s)	Tallow—Temp. (°C)	Veg. Short.—Temp. (°C)	Coconut Oil—Temp. (°C)
0.	50.	50.	40.	165	32	29	20.
15	45	48	35	180.	31	28	20.
30.	40.	44	28	195	29	28	20.
45	38	41	26	210.	28	27	20.
60.	37	39	24	225	27	27	20.
75	36	36	22	240.	27	26	20.
90.	35	34	22	255	26	26	20.
105	35	32	21	270.	25	25	20.
120.	35	31	21	285	25	25	20.
135	34	31	20.	300.	25	25	20.
150.	33	30.	20.				

Approximate melting points
Coconut oil: 21°C–25°C
Tallow: 35°C–42°C
Lard: 40°C–46°C
Vegetable shortening: 31°C–38°C
(Because of the variable nature of these mix-tures, these melting points cannot be provided more precisely.)
Students will probably recommend coconut oil because it takes less heat energy to reach its melting point and less heat energy to actually melt.
Students should note some of the following possible sources of error in their final reports: it was hard to get the solid samples to the bot-tom of the test tubes without leaving some be-hind; the cooling of the glass of the test tube and the small amount of warming of the water in the cool-water bath were not taken into ac-count; and it was difficult to identify an exact melting point for the solid.

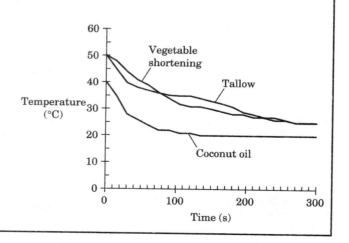

Heat of Fusion

Objectives

Students will
- use appropriate lab safety procedures.
- measure the volume of a sample of water with a graduated cylinder.
- calculate the mass of water using density.
- calculate the number of moles of water using molar mass.
- measure temperature with a thermometer or LEAP System with thermistor probe.
- use a calorimeter to measure the changes in temperature that occur as ice melts.
- apply the specific heat capacity to relate temperature changes to enthalpy changes.
- calculate the molar heat of fusion of ice.

Planning

Recommended Time

1 lab period

Materials

(for each lab group)
- Distilled water
- Ice, about 200 g total for three trials
- 400 mL beakers, 2
- 100 mL graduated cylinder
- Balance
- Beaker tongs
- Calorimeter from Exploration 9-1, or two plastic foam cups
- Glass stirring rod
- Ring stand

Bunsen burner option
- Bunsen burner and related equipment
- Ring clamp
- Wire gauze with ceramic center

Hot plate option
- Hot plate

LEAP System option
- LEAP System
- LEAP thermistor probe

Thermometer option
- Thermometer, nonmercury
- Thermometer clamp

Solution/Material Preparation

1. The amounts of ice given are merely suggestions. About three or four ice cubes from a typical ice-cube tray will provide the amount recommended for this Exploration. Larger amounts may be used, pro-

Required Precautions

- Goggles and a lab apron must be worn at all times to provide protection for the eyes and clothing.
- Tie back long hair and loose clothing.
- Read all safety cautions, and discuss them with your students.
- Remind students to use beaker tongs when handling the beaker containing hot water because it can burn or scald.
- If the LEAP System with thermistor probe or a hot plate is used, the precautions listed in the teacher's notes on page T37 must be followed to avoid electric shock.

vided that they will fit into a calorimeter. Smaller amounts are acceptable, but the results may not be as accurate. Either dispense the ice in 250 mL beakers, or allow students to obtain it themselves.

2. Reusing the calorimeters made for Exploration 9-1 can save a lot of time when students perform this Exploration.

Student Orientation

Techniques to Demonstrate

At this point in the year, students should be familiar with all of the equipment necessary for this lab. Be sure to emphasize that thermometers are too fragile to be used as stirring rods.

Make certain students understand that the more airtight their calorimeter is, the less heat energy will escape. Demonstrate how to quickly transfer the ice into the calorimeter without splashing water out of the calorimeter.

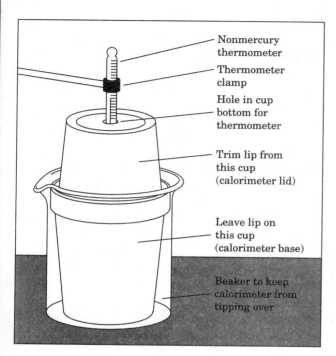

- Nonmercury thermometer
- Thermometer clamp
- Hole in cup bottom for thermometer
- Trim lip from this cup (calorimeter lid)
- Leave lip on this cup (calorimeter base)
- Beaker to keep calorimeter from tipping over

Pre-Lab Discussion

Be certain that students are familiar with the specific heat capacity equation, calorimeter measurements and calculations, and the concepts, such as the law of conservation of energy, that underlie calorimetry techniques.

Sample Data

Measurement	Trial 1	Trial 2	Trial 3
Init. H_2O vol. (mL)	100.0	98.5	99.0
Init. H_2O temp. (°C)	50.0	50.0	50.0
Final H_2O vol. (mL)	161.0	159.2	160.0
Final H_2O temp. (°C)	1.0	1.1	1.0

Post-Lab

Disposal

All water and ice may be poured down the drain.

Answers to

Analysis and Interpretation

1. 61.0 mL ice for *Trial 1*
   60.7 mL ice for *Trial 2*
   61.0 mL ice for *Trial 3*

2. 61.0 g ice for Trial 1, 60.7 g ice for Trial 2, 61.0 g ice for Trial 3

3. $61.0 \text{ g} \times \dfrac{1 \text{ mol } H_2O}{18.02 \text{ g}} =$

   3.39 mol ice for Trial 1

   $60.7 \text{ g} \times \dfrac{1 \text{ mol } H_2O}{18.02 \text{ g}} =$

   3.37 mol ice for Trial 2

   $61.0 \text{ g} \times \dfrac{1 \text{ mol } H_2O}{18.02 \text{ g}} =$

   3.39 mol ice for Trial 3

4. $100.0 \text{ mL} \times \dfrac{0.988 \text{ g}}{1 \text{ mL}} =$

   98.8 g H_2O for Trial 1

   $98.5 \text{ mL} \times \dfrac{0.988 \text{ g}}{1 \text{ mL}} =$

   97.3 g H_2O for Trial 2

   $99.0 \text{ mL} \times \dfrac{0.988 \text{ g}}{1 \text{ mL}} =$

   97.8 g H_2O for Trial 3

5. $\Delta t_{warm,1} = 49.0°C$
   $\Delta t_{warm,2} = 48.9°C$
   $\Delta t_{warm,3} = 49.0°C$

6. $\Delta t_{cool,1} = 1.0°C$
   $\Delta t_{cool,2} = 1.1°C$
   $\Delta t_{cool,3} = 1.0°C$

Conclusions

7. (energy absorbed by ice) + (energy absorbed by water from melted ice) = (energy released by warm water) Equating these quantities is possible because, according to the law of conservation of energy, energy cannot be created or destroyed, only changed from one form to another or transferred from one object to another.

8. The energy absorbed by the water from the melted ice and the energy released by the warm water can be calculated using the Δt values, the mass values, and the specific heat capacity of water.

9. **Trial 1**
heat released by warm water:
$(98.8 \text{ g})(49.0°\text{C})(4.184 \text{ J/g} \cdot °\text{C}) = 20.3 \text{ kJ}$
heat absorbed by water from melted ice: $(61.0 \text{ g})(1.0°\text{C})(4.184 \text{ J/g} \cdot °\text{C}) =$
0.26 kJ

Trial 2
heat released by warm water:
$(97.3 \text{ g})(48.9°\text{C})(4.184 \text{ J/g} \cdot °\text{C}) = 19.9 \text{ kJ}$
heat absorbed by water from melted ice: $(60.7 \text{ g})(1.1°\text{C})(4.184 \text{ J/g} \cdot °\text{C}) =$
0.28 kJ

Trial 3
heat released by warm water:
$(97.8 \text{ g})(49.0°\text{C})(4.184 \text{ J/g} \cdot °\text{C}) = 20.1 \text{ kJ}$
heat absorbed by water from melted ice: $(61.0 \text{ g})(1.0°\text{C})(4.184 \text{ J/g} \cdot °\text{C}) =$
0.26 kJ

10. heat absorbed by ice, Trial 1: 20.3 kJ − 0.26 kJ = 20.0 kJ
heat absorbed by ice, Trial 2: 19.9 kJ − 0.28 kJ = 19.6 kJ
heat absorbed by ice, Trial 3: 20.1 kJ − 0.26 kJ = 19.8 kJ

11. $\dfrac{20.0 \text{ kJ}}{61.0 \text{ g ice}} = 0.328 \text{ kJ/g ice}$

$\dfrac{19.6 \text{ kJ}}{60.7 \text{ g ice}} = 0.323 \text{ kJ/g ice}$

$\dfrac{19.8 \text{ kJ}}{61.0 \text{ g ice}} = 0.325 \text{ kJ/g ice}$

$\text{average} = \dfrac{0.328 + 0.323 + 0.325}{3} =$
0.325 kJ/g ice

12. $\dfrac{0.325 \text{ kJ}}{1 \text{ g ice}} \times \dfrac{18.02 \text{ g ice}}{1 \text{ mol}} = 5.86 \text{ kJ/mol}$

13. $2500 \text{ cm} \times 6000 \text{ cm} \times 2.5 \text{ cm} \times$
$\dfrac{0.917 \text{ g}}{1 \text{ mL}} \times \dfrac{0.325 \text{ kJ}}{1 \text{ g}} = 1.1 \times 10^7 \text{ kJ}$

Extensions

1. The accepted value for the heat of fusion of water is 6.02 kJ/mol.

$\dfrac{5.86 - 6.02}{6.02} \times 100 = 2.7\% \text{ error}$

2. Students' answers will vary. For most students, the calibration factor will be so small as to be insignificant compared to the magnitude of the values in this lab and the precision to which these values are measured.

3. Students' answers will vary. Students may suggest improving the technique by measuring masses of water and ice instead of volume. This can be done by measuring the mass of the calorimeter before filling, after filling with warm water, and after the ice has melted in it. Be certain student answers are safe and include carefully planned procedures.

4. Even though the outside temperature was well below 0°C, the temperature in the cold cellar remained above the freezing point of water because heat is released by the liquid water as it freezes. The temperature of the room could not drop below 0°C until all of the water in the barrels had frozen.

Heat of Solution

Objectives

Students will

- use appropriate lab safety procedures.
- measure the temperature change when a salt dissolves in a calorimeter using a thermometer or a LEAP System with thermistor probe.
- measure mass using a laboratory balance.
- relate temperature changes in a calorimeter to changes in heat energy.
- determine the amount of heat energy released in joules for each gram of sodium thiosulfate pentahydrate, $Na_2S_2O_3 \cdot 5H_2O$, that dissolves.
- determine the amount of heat energy released in joules for each gram of sodium acetate, $NaC_2H_3O_2$, that dissolves.
- evaluate the cost effectiveness of $Na_2S_2O_3 \cdot 5H_2O$ versus $NaC_2H_3O_2$ by comparing their costs per gram.
- evaluate their lab technique for precision.

Planning

Recommended Time

1 lab period

Materials

(for each lab group)
- Ice
- $NaC_2H_3O_2$, 30 g
- $Na_2S_2O_3 \cdot 5H_2O$, 30 g
- 600 mL beaker
- 100 mL graduated cylinder
- Balance, centigram
- Chemical references and supply catalogs
- Corrugated cardboard or lid for plastic foam cup
- Pencil with point
- Plastic washtub
- Plastic foam cup, small
- Test tubes, Pyrex, small, 4
- Test-tube rack
- Wire, 15 cm

LEAP System option
- LEAP System
- LEAP thermistor probe

Thermometer option
- Thermometer, nonmercury

Solution/Material Preparation

1. Put out chemical supply catalogs as references for pricing $NaC_2H_3O_2$ and $Na_2S_2O_3 \cdot 5H_2O$.

Required Precautions

- Goggles, gloves, and a lab apron must be worn at all times.
- Read all safety cautions, and discuss them with your students.
- Lids should remain on reagent bottles when students are not using them. The $Na_2S_2O_3 \cdot 5H_2O$ can lose water to warm, dry air or absorb water from moist air.
- If the LEAP System with thermistor probe is used, the precautions listed in the teacher's notes on page T37 must be followed to avoid electric shock.

2. To save time, the complete thermometer/thermistor assemblies can be prepared in advance. Cut 15 cm lengths of wire. (Any gauge will do, provided it is easily bent.) Insert the length into one of the holes in the cardboard lid. At each end make a loop that has a diameter of about 1 cm. Bend the wire at the loops so that each loop is perpendicular to the rest of the wire.

3. Another way to save time is to reuse the calorimeters from Exploration 9-1, making another hole for the stirrer.

Student Orientation

Because students often interpret dumping the salt from the test tube as pouring or adding in portions, they may benefit from a demonstration of how to carefully dump the salt from the test tube all at once. Remind them to pour the salt near the edge of the cup and not onto the thermometer to avoid false peak temperatures. If the salt sticks to the side of the cup, a quick swirling motion can agitate enough water to wash the salt back into the cup.

Demonstrate the use and purpose of the stirrer. Stirring usually implies a circular motion, but here experimental design demands an up-and-down motion of the stirrer. Although this may be perceived as awkward and may be forgotten altogether, other methods cause the thermometer to be knocked about, possibly breaking it.

Pre-Lab Discussion

Remind students that distilled water, not tap water, is to be used because the water participates in the reaction. It is not a bath. The peak point of the temperature should be reached in 5 min or less. Students need to know that the quality of their technique is the focus of this experiment.

Because the solids are ionic and contain both a monatomic ion and a polyatomic ion, this is a good opportunity to review identification and naming of ions. Review the formation of ionic crystals in Section 5-2 of the text and the summation of heat energies in Figure 5-10. Briefly present the dissolution process and summation of the heat energies involved. Review percent error, percent difference, standard deviation, and their relationships to accuracy and precision.

Post-Lab

Disposal

Set out two disposal containers. One is for sodium thiosulfate pentahydrate, $Na_2S_2O_3 \cdot 5H_2O$, and its solutions. The other is for sodium acetate, $NaC_2H_3O_2$, and its solutions. Allow the excess water to evaporate from the solutions, and store the solid salts in labeled reagent bottles for reuse. It will be necessary to pulverize the crystals into smaller chunks before reusing them.

Answers to

Analysis and Interpretation

1. $\Delta t_1 = 23.5°C - 13.5°C = 10.0°C$
   $\Delta t_2 = 23.9°C - 14.0°C = 9.9°C$
   $\Delta t_3 = 24.5°C - 14.5°C = 10.0°C$
   $\Delta t_4 = 24.9°C - 15.0°C = 9.9°C$

2. Heat absorbed $= m_{H_2O} \times c_{p, H_2O} \times \Delta t$
   $D_{H_2O} = 1.00$ g/mL;

   75.0 mL $\times D_{H_2O} = 75.0$ g

   Trial 1: 75.0 g $\times \dfrac{4.180 \text{ J}}{\text{g} \cdot °C} \times 10.0°C =$
   $$3.14 \times 10^3 \text{ J}$$

   Trial 2: 75.0 g $\times \dfrac{4.180 \text{ J}}{\text{g} \cdot °C} \times 9.9°C =$
   $$3.10 \times 10^3 \text{ J}$$

   Trial 3: 75.0 g $\times \dfrac{4.180 \text{ J}}{\text{g} \cdot °C} \times 10.0°C =$
   $$3.14 \times 10^3 \text{ J}$$

   Trial 4: 75.0 g $\times \dfrac{4.180 \text{ J}}{\text{g} \cdot °C} \times 9.9°C =$
   $$3.10 \times 10^3 \text{ J}$$

3. Trial 1: $\dfrac{3140 \text{ J}}{15.0 \text{ g}} =$
   $$2.09 \times 10^2 \text{J/g } Na_2S_2O_3 \cdot 5H_2O$$

Sample Data and Analysis

Measurement	Trial 1—$Na_2S_2O_3 \cdot 5H_2O$	Trial 2—$Na_2S_2O_3 \cdot 5H_2O$	Trial 3—$NaC_2H_3O_2$	Trial 4—$NaC_2H_3O_2$
Mass of solute (g)	15.0	15.0	15.0	15.2
Volume of cold H_2O (mL)	75.0	75.0	75.0	75.0
Initial H_2O temp. (°C)	13.5	14.0	14.5	15.0
Final H_2O temp. (°C)	23.5	23.9	24.5	24.9

$$\text{Trial 2: } \frac{3100 \text{ J}}{15.0 \text{ g}} =$$

$$2.07 \times 10^2 \text{ J/g Na}_2\text{S}_2\text{O}_3 \cdot 5\text{H}_2\text{O}$$

4. $$\text{Trial 3: } \frac{3140 \text{ J}}{15.0 \text{ g}} =$$

$$2.09 \times 10^2 \text{ J/g NaC}_2\text{H}_3\text{O}_2$$

$$\text{Trial 4: } \frac{3100 \text{ J}}{15.2 \text{ g}} =$$

$$2.04 \times 10^2 \text{ J/g NaC}_2\text{H}_3\text{O}_2$$

5. $$\frac{2.09 \times 10^2 \text{ J/g} - 2.000 \times 10^2 \text{ J/g}}{2.000 \times 10^2 \text{ J/g}} \times$$

$100 = 4.5\%$ error

$$\frac{2.07 \times 10^2 \text{ J/g} - 2.000 \times 10^2 \text{ J/g}}{2.000 \times 10^2 \text{ J/g}} \times$$

$100 = 3.5\%$ error

6. average =
$$2.08 \times 10^2 \text{ J/g Na}_2\text{S}_2\text{O}_3 \cdot 5\text{H}_2\text{O}$$
average $= 2.06 \times 10^2 \text{ J/g NaC}_2\text{H}_3\text{O}_2$
$$\frac{2.09 \times 10^2 \text{ J/g} - 2.08 \times 10^2 \text{ J/g}}{2.08 \times 10^2 \text{ J/g}} \times$$
$100 = 0.5\%$ difference for Trial 1
0.5% difference for Trial 2
$$\frac{2.09 \times 10^2 \text{ J/g} - 2.06 \times 10^2 \text{ J/g}}{2.06 \times 10^2 \text{ J/g}} \times$$
$100 = 1.5\%$ difference for Trial 3
1.0% difference for Trial 4
In all cases, the sample data is reasonably precise.

7. Cost of $\text{Na}_2\text{S}_2\text{O}_3 \cdot 5\text{H}_2\text{O}$: $0.0085/g
Cost of $\text{NaC}_2\text{H}_3\text{O}_2$: $0.0242/g
These values were calculated from a chemical supply catalog, which offered these chemicals in practical grade prices of $8.50/kg and $12.10/500 g, respectively. Students' discussions should indicate that thiosulfate appears more cost-effective because nearly the same amount of energy per gram is provided at a much cheaper cost.

8. The plastic foam is an insulating material, so very little heat is transferred to the calorimeter or the surroundings outside the calorimeter. Glass is a poorer insulator than plastic foam is; more heat is lost to the calorimeter and surroundings.

9. Students' suggestions for improving the calorimeter apparatus will vary, but may include using a thermometer sealed into the cup's lid, using a magnetic stir bar within the cup, using a plastic foam lid with grooves for the rim of the cup, or setting the cup inside a second plastic foam cup.

Conclusions

10. The ions lose potential energy as they become surrounded by water molecules. This energy is transferred to the water or the surroundings.

11. The change in enthalpy for crystallization would be the same as the change in enthalpy for dissolution, probably about 200 J/g, or 16 kJ/mol.

Extensions

1. A steady, rather than fluctuating, amount of heat is entering the surroundings. This reduces the risk of the temperature rising too high and the user being burned.

2.

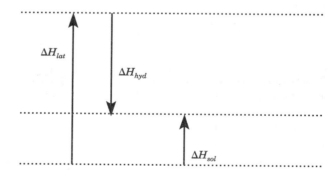

Gas Pressure-Volume Relationship

10-1

Technique Builder

Students will

- use appropriate lab safety procedures.
- measure mass using a laboratory balance.
- measure length using a metric ruler.
- measure volume using a syringe or Boyle's law apparatus.
- measure volume of a gas as pressure is increased.
- calculate the pressure exerted on a gas given mass, acceleration due to gravity, and area.
- graph the pressure-volume data in two ways.
- express the pressure-volume relationship mathematically.
- calculate the amount of gas contained in a tank at different pressures.
- evaluate the different possible unit costs of tanks made with valves tested to withstand different pressures.

Planning

Recommended Time

1 lab period

Materials

(for each lab group)
- Balance
- Carpet thread
- Removable Post-It notes
- Ruler metric
- Sealed syringe or Boyle's law apparatus
- Stackable objects of roughly equal mass, such as books or blocks of wood (500–550 g each), 4

Solution/Material Preparation

I. Boyle's law apparatus can be purchased from some scientific supply houses. Supply houses that don't offer them often sell syringes that are permanently sealed to use instead. The syringes typically have a smaller volume, which means that the results may not be as accurate. Whichever option you choose, you may need to adapt the equipment to be sure that the masses can be stacked on it. Do not use any glass equipment, because it can break and cause cuts. The apparatus or syringe should be made from plastic. **Do not, under any circumstance, use any older "Boyle's law" apparatus that contains columns of mercury.** The hazards related to the chronic release of toxic

Required Precautions

- Always wear goggles and a lab apron to provide protection for eyes and clothing. Even though there are no hazardous chemicals used in this particular procedure, there could be hazardous chemicals and equipment in the lab.
- Read all safety cautions, and discuss them with your students.
- Thoroughly discuss the potential hazard of stacked books or blocks falling off the apparatus. Each mass should be placed on the stack with care, and one member of the team should keep a hand on either side of the stack, ready to stabilize it if it wavers. The other team member should not attempt to read the volume until a team member is ready to catch the masses.

mercury vapor into the air and the possibility of a mercury spill do not balance the few benefits obtained from using such equipment.

2. Each stackable object should be approximately 500 g. Less mass provides too small a change in volume. If masses in excess of 600 g each are used, the pres-

sure will be so excessive that air will escape around the gasket.

Student Orientation

Review the general equation for a straight line and the meaning of proportionality and inverse proportionality.

Techniques to Demonstrate
Show students how to use the Boyle's law apparatus, especially how to twist the piston to reduce frictional forces. The frictional force is quite significant, but if the piston is twisted until a stable volume is reached, reliable data can be obtained. Warn students that the apparatus must not have any leaks. If the apparatus leaks, the piston will not return to its original position regardless of the number of times the head is twisted. The most probable source of leakage is the end of the syringe where the needle normally is attached.

Post-Lab

Disposal
No disposal required.

Answers to

Analysis and Interpretation

1. 0 mass: 0 g
   1 mass: 499 g
   2 masses: 1001 g
   3 masses: 1509 g
   4 masses: 2016 g

2. $A = \pi\left(\dfrac{36 \times 10^{-3}\ m}{2}\right)^2 = 1.0 \times 10^{-3}\ m^2$

Sample Data

Masses added	Mass (g)	Trial 1 vol. (cm³)	Trial 2 vol. (cm³)	Trial 3 vol. (cm³)
0 mass	0	35.0	35.0	35.0
Mass no. 1	499	33.6	33.1	33.3
Mass no. 2	502	31.9	32.3	32.0
Mass no. 3	508	30.4	30.8	30.3
Mass no. 4	507	28.9	29.2	29.4

diameter of syringe barrel: 36 mm

3. 1 mass: $0.499\ kg \times 9.801\ m/s^2 = 4.89\ N$
   2 masses: $1.001\ kg \times 9.801\ m/s^2 = 9.811\ N$
   3 masses: $1.509\ kg \times 9.801\ m/s^2 = 14.79\ N$
   4 masses: $2.016\ kg \times 9.801\ m/s^2 = 19.76\ N$

4. $P_1 = \dfrac{4.89\ N}{1.0 \times 10^{-3}\ m^2} = 4.9 \times 10^3\ Pa$

   $P_2 = \dfrac{9.811\ N}{1.0 \times 10^{-3}\ m^2} = 9.8 \times 10^3\ Pa$

   $P_3 = \dfrac{14.79\ N}{1.0 \times 10^{-3}\ m^2} = 1.5 \times 10^4\ Pa$

   $P_4 = \dfrac{19.76\ N}{1.0 \times 10^{-3}\ m^2} = 2.0 \times 10^4\ Pa$

5. The pressure on the trapped air before masses are added to the plunger is normal atmospheric pressure, $1.013 \times 10^5\ Pa$.

6. $P_0 = 1.013 \times 10^5\ Pa$
   $P_1 = 4.9 \times 10^3\ Pa + 1.013 \times 10^5\ Pa = 1.062 \times 10^5\ Pa$
   $P_2 = 9.8 \times 10^3\ Pa + 1.013 \times 10^5\ Pa = 1.111 \times 10^5\ Pa$
   $P_3 = 1.5 \times 10^4\ Pa + 1.013 \times 10^5\ Pa = 1.163 \times 10^5\ Pa$
   $P_4 = 2.0 \times 10^4\ Pa + 1.013 \times 10^5\ Pa = 1.213 \times 10^5\ Pa$

7. $V_{0(avg)} = 35.0\ cm^3$
   $V_{1(avg)} = 33.3\ cm^3$
   $V_{2(avg)} = 32.1\ cm^3$
   $V_{3(avg)} = 30.5\ cm^3$
   $V_{4(avg)} = 29.2\ cm^3$

8. As the pressure increased, volume decreased.

9.

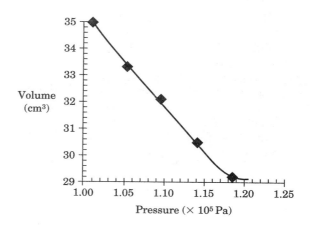

10.

$$1/V_{0(avg)} = \frac{1}{35.0 \text{ cm}^3} = 2.86 \times 10^{-2} \text{ cm}^{-3}$$

$$1/V_{1(avg)} = \frac{1}{33.3 \text{ cm}^3} = 3.00 \times 10^{-2} \text{ cm}^{-3}$$

$$1/V_{2(avg)} = \frac{1}{32.1 \text{ cm}^3} = 3.12 \times 10^{-2} \text{ cm}^{-3}$$

$$1/V_{3(avg)} = \frac{1}{30.5 \text{ cm}^3} = 3.28 \times 10^{-2} \text{ cm}^{-3}$$

$$1/V_{4(avg)} = \frac{1}{29.2 \text{ cm}^3} = 3.42 \times 10^{-2} \text{ cm}^{-3}$$

11.

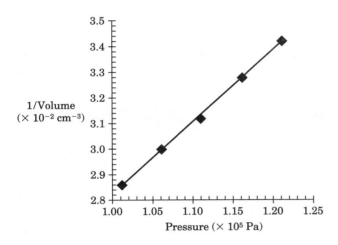

12. $1/V = (2.79 \times 10^{-7})P + 2.66 \times 10^{-4}$
(from graphics calculator)

13. The temperature did not change because any temporary increases in temperature would be dissipated throughout the room. The sealed container does not allow any additional gas particles in or out.

14. The carpet thread allows a small amount of air to escape as the plunger is set at the appropriate level. If no air were allowed to escape, you could not be certain that the initial volume was actually the volume at normal atmospheric pressure.

Conclusions

15. $1/V = (2.79 \times 10^{-7})P + 2.66 \times 10^{-4}$
Because 2.66×10^{-4} is so much smaller than the product of 2.79×10^{-7} and P, consider it to be zero.

$$1/V = (2.79 \times 10^{-7})P + 0$$

$$\frac{1}{(2.79 \times 10^{-7})} = k = PV$$

16. The same value of k must not be used because the air in the tank is a different amount and could be at a different temperature.

17. Because $PV = k$, the following relationship is also true.
$$P_{norm}V_{norm} = P_{tank}V_{tank} = k$$

18. $\dfrac{10.5 \text{ L} \times 112 \text{ atm}}{1 \text{ atm}} = 1.18 \times 10^3 \text{ L}$

of air for a PDE-57 tank

$\dfrac{10.5 \text{ L} \times 168 \text{ atm}}{1 \text{ atm}} = 1.76 \times 10^3 \text{ L}$

of air for a RRM-82 tank

$\dfrac{10.5 \text{ L} \times 219 \text{ atm}}{1 \text{ atm}} = 2.30 \times 10^3 \text{ L}$

of air for a RMD-332 tank

$\dfrac{10.5 \text{ L} \times 240 \text{ atm}}{1 \text{ atm}} = 2.52 \times 10^3 \text{ L}$

of air for a ZRH-61 tank

$\dfrac{10.5 \text{ L} \times 253 \text{ atm}}{1 \text{ atm}} = 2.66 \times 10^3 \text{ L}$

of air for a XJP-27 tank

19. $1.18 \times 10^3 \text{ L} \times \dfrac{26 \text{ min}}{1000 \text{ L}} = 31 \text{ min}$ for a
PDE-57 tank

$1.76 \times 10^3 \text{ L} \times \dfrac{26 \text{ min}}{1000 \text{ L}} = 46 \text{ min}$ for a
RRM-82 tank

$2.30 \times 10^3 \text{ L} \times \dfrac{26 \text{ min}}{1000 \text{ L}} = 60. \text{ min}$ for a
RMD-332 tank

$2.52 \times 10^3 \text{ L} \times \dfrac{26 \text{ min}}{1000 \text{ L}} = 65 \text{ min}$ for a
ZRH-61 tank

$2.66 \times 10^3 \text{ L} \times \dfrac{26 \text{ min}}{1000 \text{ L}} = 69 \text{ min}$ for a
XJP-27 tank

20. PDE-57 tank: $37.50 + $50.00 + $20.00 + $1.50 = $109.00

$\dfrac{\$109.00}{31 \text{ min}} = \$3.50/\text{min}$

RRM-82 tank: $\dfrac{\$111.50}{46 \text{ min}} = \$2.40/\text{min}$

RMD-332 tank: $\dfrac{\$116.50}{60. \text{ min}} = \$1.90/\text{min}$

(most cost-effective)

ZRH-61 tank: $\dfrac{\$141.50}{65 \text{ min}} = \$2.20/\text{min}$

XJP-27 tank: $\dfrac{\$176.50}{69 \text{ min}} = \$2.60/\text{min}$

Extensions

1. Gas pressure results from particles hitting the sides of the container. If the size of the container is decreased while the number of molecules remains the same, the number of collisions with the container walls increases.

2. In order for the plunger and masses to remain in a stable position, the gas must be exerting an upward pressure of the same magnitude as the downward pressure of the piston and masses. This pressure is exerted in all directions and is considered the pressure of the gas.

3. Student answers will vary. Some may notice that the force due to the weight of the plunger was not considered and may propose measuring it and including it in the calculations. Others may suggest multiple trials or using an apparatus that can hold larger volumes that are easier to read.

4. At an infinitely high pressure, the volume should be infinitely small, approaching zero.

5. Be sure students' work indicates a recognition that significant discoveries are sometimes made with very simple apparatus. An appreciation of Boyle's apparatus can help students understand how pressures are measured with manometers.

Additional Notes

Gas Temperature-Volume Relationship Balloon Flight

INVESTIGATION 10-1

Problem Solving

Objectives

Students will
- use appropriate lab safety procedures.
- measure temperature using a thermometer or LEAP System thermistor probe.
- measure volume of a sample of gas at different temperatures.
- convert temperatures in degrees Celsius to kelvins.
- graph temperature in kelvins versus volume.
- use the graph to relate any volume and temperature.
- compare results to Charles's law.
- calculate the mass of a gas given density.
- use results of procedure to calculate the temperature and volume necessary to keep a hot-air balloon aloft.

Planning

Recommended Time
1 lab period

Prerequisite
Exploration 10-1

Materials
(for each lab group)
- Beaker tongs
- Lubricant (mineral oil or glycerin)
- 400 mL beaker
- 10 cm³ sealed syringe
- Clamps, 2

Bunsen burner option
- Bunsen burner, gas tubing, striker
- Ring
- Ring stand
- Wire gauze

Hot plate option
- Hot plate
- Ring stand

LEAP System option
- LEAP thermistor probe

Thermometer option
- Thermometer, nonmercury
- Rubber stopper, one-hole, split

Estimated cost of materials:
$39 000–$44 000

Required Precautions

- Goggles and a lab apron must be worn at all times.
- Loose hair and clothing must be tied back.
- Remind students that hot equipment doesn't always look hot, so they should use beaker tongs to handle glassware and other equipment.
- Read all safety cautions, and discuss them with your students.
- If the LEAP System with thermistor probe is used, the precautions listed in the teacher's notes on page T37 must be followed to avoid electric shock.

Solution/Material Preparation

1. Either glycerin or mineral oil may be used as a lubricant. The advantage of glycerin is that it is water-soluble and easier to clean up.
2. The syringe from the Boyle's law apparatus can be adapted for this experiment.

Temperature (°C/K)	Volume of air (cm³)
20/293	5.0
40/313	5.2
60/333	5.6
80/353	6.0
100/373	6.4

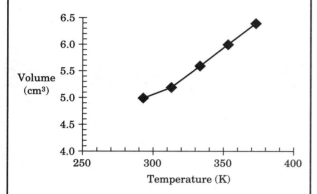

A linear relationship is involved.

$$T = kV \quad \text{and} \quad \frac{V_1}{T_1} = \frac{V_2}{T_2}$$

The only point off the line is at 293 K, which would be predicted to be 4.8 cm³ instead of 5.0 cm³.

$$\frac{5.0 \text{ cm}^3 - 4.8 \text{ cm}^3}{5.0 \text{ cm}^3} = 4\% \text{ error}$$

mass of air inside balloon:

$$(2.0 \times 10^3 \text{ m}^3) \times \frac{1000 \text{ L}}{1 \text{ m}^3} \times$$

$$\frac{1.22 \text{ g of air at start}}{1 \text{ L}} = 2.4 \times 10^6 \text{ g}$$

total mass:
$(2.4 \times 10^6 \text{ g air}) + (4.8 \times 10^5 \text{ g equipment}) = 2.9 \times 10^6 \text{ g}$

The same total mass of air at flying altitude must be displaced for the balloon to float.

volume of air displaced:

$$(2.9 \times 10^3 \text{ m}^3) \times \frac{1 \text{ L}}{1.17 \text{ g}} \times \frac{1 \text{ m}^3}{1000 \text{ L}} =$$

$$2.5 \times 10^3 \text{ m}^3$$

temperature required to inflate the balloon:

$$(2.5 \times 10^3 \text{ m}^3) \times \frac{288 \text{ K}}{2.0 \times 10^3 \text{ m}^3} = 360 \text{ K},$$

$$\text{or } 87°C$$

accounting for 4% error, 87°C ± 3.5°C

However, it may be best to adjust the initial volume to one-half of the syringe's total capacity.

3. Before the lab, prepare the thermometers for student use by splitting the side of a one-hole rubber stopper. Lubricate the thermometer with a small amount of glycerin or mineral oil. Then, after putting on cloth gloves and wrapping the thermometer in cloth, you can insert the thermometer into the crack in the stopper so that it will not break when the students clamp it.

Student Orientation

Pre-Lab Discussion

Students should not expect exact results in this experiment. Instead they should be looking for trends and reproducible data. The data will establish the relationship between volume and temperature and verify Charles's law.

Discuss the need to make measurements of volume at several different temperatures, ranging from room temperature to boiling (100°C). Encourage students to repeat the procedure to verify results.

Discuss with students the mechanism by which a balloon stays afloat. The balloon will sink when the mass of the balloon and the hot air it contains exceeds the mass of a similar volume of the surrounding air.

Techniques to Demonstrate

Demonstrate how to lubricate the rubber seal on the syringe plunger with a little mineral oil so that the plunger can move easily in the barrel of the syringe. Emphasize that the syringe and thermometer should be clamped so that the entire scale on each is visible. Be sure students know to heat the water bath gently.

Tips for Evaluating the Pre-Lab Requirements

Be certain that proposed procedures recognize all necessary safety precautions, especially those involved with the use of the Bunsen burner. Be certain student data tables indicate that they will measure the volume of the gas for at least four different temperatures ranging from room temperature to 100°C. (The more temperatures measured, the better the results are likely to be, but the constraints of lab time must be considered by students as thay make their plans.)

Proposed Procedure

Lubricate the plunger of a sealed syringe with a small amount of mineral oil or glycerin. Fill the syringe with 5.0 cm³ of air. Clamp the syringe into a hot-water bath. Record the exact temperature of the water using a thermometer or LEAP System thermistor probe and the initial volume. Record temperature and volume measurements at 20°C intervals until 100°C is reached. Then graph the data to determine the relationship between temperature and volume and the magnitude of any deviations from ideal gas behaviors. Sample calculations for relating these results to the problem of the balloon are given in the Sample Data and Analysis section.

Post-Lab

Disposal

There are no special disposal requirements, but students should thoroughly clean the plungers of the syringes with soap and water after completing the procedure.

Additional Notes

Masses of Equal Volumes of Gases

EXPLORATION

10-2

Technique Builder

Objectives

Students will
- use appropriate lab safety procedures.
- collect samples of gas at room temperature and pressure.
- measure the masses of the samples using a laboratory balance.
- measure the volume of the gases by water displacement.
- determine the actual masses of the gases by accounting for buoyancy.
- calculate the ratio of the two masses.
- apply Avogadro's principle to determine the identity of gases by determining the ratio of their molar masses.

Planning

Recommended Time

1 lab period

Materials

(for each lab group)
- Dry ice
- Methane gas (from gas jet)
- 2 L glass bottle
- 250 mL or 500 mL graduated cylinder
- Balance
- Crucible tongs
- Medicine dropper
- Pinch clamp
- 1 L plastic bag
- Plastic straw
- Pneumatic trough
- Rubber band
- Rubber stopper, one-hole, no. 6
- Rubber tubing, 50 cm
- Tongs

Optional equipment
- Barometer

Solution/Material Preparation

1. Each lab group only needs a small, marble-sized piece of dry ice.
2. Before the lab, prepare several sets of medicine droppers for use. First, put on cloth gloves and wrap the medicine droppers in a layer of cloth. Insert them into the one-hole rubber stoppers. This way,

Required Precautions

- Goggles and a lab apron must be worn at all times.
- Read all safety cautions, and discuss them with your students.
- Remind students not to touch the dry ice with their hands.
- Make sure that the fume hood is turned on and working properly before starting the lab. Methane should be expelled only in the hood or outdoors.
- Be certain that the room is well-ventilated before starting the lab.
- Methane is flammable; mixtures of air and methane containing more than 5% and less than 15% methane will explode if ignited. Make sure there are no flames, sparks, or other sources of ignition in the lab. Thoroughly ventilate the room before it is used for another lab. (Note: if the directions are followed, the total quantity of methane used by the students will be less than 5% of the total air in the room.)

the medicine dropper will not break when the bag is sealed with a rubber band.
3. If a barometer is not used, listen to a weather report to determine air pressure in the lab.

Student Orientation

Pre-Lab Discussion

Students may need to discuss the meaning of apparent and actual mass of a buoyant object. Use the example of the buoyant effect of water on a swimmer or a boat to help clarify the concept.

Review the use of the pressure-measuring equipment available in your lab and the reason why atmospheric pressure is needed in this experiment.

Discuss the importance of releasing the methane in the hood.

Techniques to Demonstrate

Demonstrate how to assemble the bag assembly for collecting the gases. Show students a bag assembly that is filled with air to a pressure greater than ambient pressure, and then allow the pressure in the bag to equilibrate with that of the atmosphere. If students are unfamiliar with the use of a pneumatic trough, demonstrate the placement and removal of the bottle and the rubber tubing.

Post-Lab

Disposal

No disposal is required.

Answers to

Analysis and Interpretation

1. The masses were apparent masses because the volume of the gas displaced a certain amount of air which buoyed up the sample slightly.

2. 29.95 g − 27.30 g = 2.65 g apparent mass of CO_2
   27.36 g − 27.30 g = 0.06 g apparent mass of CH_4

3. For the conditions in the sample data, the density of air is 1.17 g/L.

Sample Data

Sample Data	
Lab temperature	25°C
Atmospheric pressure	750.0 torr
Mass of empty bag assembly	27.30 g
Mass of bag assembly and CO_2	29.95 g
Mass of bag assembly and CH_4	27.36 g
Volume of bag	1230 mL

$$1230 \text{ mL} \times \frac{1 \text{ L}}{1000 \text{ mL}} \times \frac{1.17 \text{ g}}{1 \text{ L}} =$$

1.44 g air displaced by gas samples

4. Actual mass of CO_2 = 1.44 g + 2.65 g = 4.09 g CO_2
   Actual mass of CH_4 = 1.44 g + 0.06 g = 1.50 g CH_4

5. The ratio of actual masses is:
$$\frac{4.09 \text{ g } CO_2}{1.50 \text{ g } CH_4} = 2.73 : 1.00$$

Conclusions

6. $\dfrac{44.01 \text{ g/mol } CO_2}{16.05 \text{ g/mol } CH_4} = 2.74 : 1.00$

7. Student errors will range from 0% to 15%. For the sample data:
$$\frac{2.74 - 2.73}{2.74} = 0.4\%$$

8. Based on these calculations, it is unlikely that the tanks contain CO_2 and CH_4 because the 2.74:1.00 ratio is far from the 2.00:1.00 ratio.

9. $\dfrac{2.02 \text{ g } H_2}{16.05 \text{ g } CH_4} = 0.126 \text{ g } H_2 : 1.00 \text{ g } CH_4;$

   7.95 g CH_4:1.00 g H_2
   15.8 g O_2:1.00 g H_2
   21.8 g CO_2:1.00 g H_2
   1.994 g O_2:1.00 g CH_4
   1.375 g CO_2:1.000 g O_2

10. The 2.00:1.00 ratio determined by Compressed Gases, Inc., is closest to the O_2:CH_4 ratio.

Extensions

1. It was important to measure the mass of methane and carbon dioxide under the same conditions of temperature and pressure because only gases at the same temperature and pressure have the same number of molecules.

2. Possible sources of error are incomplete filling of the bag, failure to equilibrate the pressure of the gas in the bag to room pressure, inaccurate measurement of the mass of the bag assembly, or a leak. Some students may also realize that the gas from the gas jet is not pure methane.

3. $\dfrac{1.30 \text{ g gas}}{1.48 \text{ g } O_2} = \dfrac{x}{32.00 \text{ g } O_2}$
   $x = 28.1$ g/mol

Paper Chromatography

Objectives

Students will

- use appropriate lab safety procedures.
- demonstrate proficiency in qualitatively separating mixtures using paper chromatography.
- evaluate samples to establish which pen was used to sign a document.
- relate the results of a chromatogram to the solubility properties of the components of a mixture.

Planning

Recommended Time

1 lab period

Materials

(for each lab group)

- Distilled water
- Isopropanol, 25 mL
- Filter papers, 12 cm, 4
- Filter paper wicks, 2 cm equilateral triangles, 2
- Forged signature on paper
- Paper clips, 2
- Pens, black ink, 6 different types
- Pencil
- Petri dish with lid
- Ruler

Solution/Material Preparation

1. Label the pens *1–6*.
2. Prepare the signatures before students enter the room. First cut a piece of filter paper in half. On each half, about 1.0 cm away from the flat edge, sign "A. Lincoln" with one of the six black ink pens. Be sure to keep track of which pen you use. Note that students will achieve best results if they use two segments of the signature that include fairly flat lines, rather than using the entire signature, or parts with vertical lines or loops.

Required Precautions

- Goggles and a lab apron must be worn at all times.
- Read all safety cautions, and discuss them with your students.
- The isopropanol is extremely flammable. It should be kept in a closed bottle in an operating fume hood. Place only 300 mL at a time in the bottle. Students should replace the lid when they are finished.
- No burners, flames, hot plates, or other heat sources should be in use in the lab when isopropanol is being used.

Student Orientation

Techniques to Demonstrate

Students may need to see an actual chromatography setup before they understand exactly what they will be trying to achieve. This is especially true for the chromatogram of the forged signature. Students need to avoid parts of the signature with loops or vertical lines because these will not make as clear a chromatogram as dots and horizontal lines.

Pre-Lab Discussion

Although this is a popular experiment for students of all ages, for it to be instructionally useful, students must understand that the solubility properties discussed in Chapter 11 can explain, in part, *why* this particular technique works. Make certain students understand this link. It is important that students realize that the writing sample may take slightly longer to form a chromatogram than did the six dots placed on the filter paper during the lab. If students make measurements of their chromatograms (how far components traveled in what time), this can be made into a more quantitative exercise.

Disposal

Set out a disposal container for any isopropanol left over at the end of the procedure. Dilute it with 10 times as much water, and pour it down the drain. Students are instructed to pour the water down the drain. The chromatograms may be discarded in the trash can.

Answers to

Analysis and Interpretation

1.

Water   Isopropanol

2. Water is a polar molecule that will dissolve polar and charged substances. Isopropanol has a small polar region but is mostly nonpolar, so it will dissolve nonpolar substances.

3. The more soluble a component is, the farther it will travel with the solvent.

4. Larger molecules are likely to move more slowly through the solvent and do not travel as far as small molecules.

5. The graphite in pencil lead is composed of sheets of covalently bonded molecules that are nonpolar and insoluble in water. Thus, they will not contaminate the chromatogram.

6. If the process continued overnight, the solvent would reach the edge of the filter paper and begin to evaporate. The slower components would catch up with the faster ones, and they would all end up near the edge of the filter paper.

7. When the ink was dipped into the solvent, most of it dissolved in the solvent in the dish instead of being pulled with the solvent across the filter paper.

Conclusions

8. The component that travels the farthest in the water chromatogram is likely to be small and ionic or polar. The component that travels the farthest in the isopropanol chromatogram is also likely to be small, but probably nonpolar.

9. Student answers will vary, depending on which pen you chose for the forged signature. Students should justify their choices by pointing out specific similarities between the forgery chromatograms and the chromatograms of the pen they chose.

Extensions

1. Student suggestions for improving the separation will vary. Some may suggest using a longer piece of filter paper or trying other solvents. Be certain answers are safe and include carefully planned procedures before allowing students to proceed.

2.

Water	Isopropanol
aspartic acid	glycine (or valine)
glycine	valine (or glycine)
valine	phenylalanine
phenylalanine	aspartic acid

The charge on aspartic acid will make it very soluble in water. The next most soluble is likely to be glycine because it is also mostly polar. The other amino acids are both mostly nonpolar, so size should be the dominant factor. In isopropanol, depending on whether size or polarity is the most important factor, glycine or valine will go first. Phenylalanine will dissolve well because it is nonpolar, but it is so much larger that it is likely to lag behind the first two. Lastly, the charges on aspartic acid will decrease its solubility in the nonpolar solvent.

3. Benzene is a nonpolar solvent. The many charges on betanidin make it likely to be last. Of the other two, beta-carotene is so much larger that it is likely to be second-to-last, even though delphinidin has some polar bonds.

Paper Chromatography— Forensic Investigation

INVESTIGATION
11-1

Problem Solving

Objectives

Students will
- use appropriate lab safety procedures.
- demonstrate proficiency in separating mixtures using paper chromatography.
- evaluate samples to establish which pen was used on a document.
- design and implement their own procedure.

Planning

Recommended Time

1 lab period

Prerequisite

Exploration 11-1

Materials

(for each lab group)
- Distilled water
- Isopropanol, 25 mL
- Filter paper, 12 cm in diameter, 3 pieces
- Paper clips, 2
- Pens, black, 4
- Petri dish and lid
- "Ransom note" written on filter paper

Estimated cost of materials: $63 000

Solution/Material Preparation

1. It is necessary to prepare the "ransom notes" ahead of time, but no more than two hours in advance, or the ink may become too dry to separate. Cut a piece of filter paper in half, and write the ransom note about 1.0 cm away from the flat edge. Be sure to keep track of which pen you use to write the note. Remember that students should only need a very small portion with which to conduct their test. It may be easiest to make one or two notes, and cut them up to distribute pieces to the lab groups instead of giving each group an entire note.

2. Possible text for ransom note: "I have it, but somebody wants to buy it. If you

Required Precautions

- Goggles and a lab apron must be worn at all times.
- Read all safety cautions, and discuss them with your students.
- The isopropanol is extremely flammable. It should be kept in a closed bottle in an operating fume hood. Place only 300 mL at a time in the bottle. Students should replace the lid when they are finished.
- No burners, flames, hot plates, or other heat sources should be in use in the lab when isopropanol is being used.

want it back, it will cost you $100 000. Put the money in a brown paper bag, and place it on the 50-yard line of the football field tonight! Don't even try to find me because I'm too smart to get caught."

Student Orientation

Techniques to Demonstrate

After completion of the Exploration, there should be no need to demonstrate any techniques.

Pre-Lab Discussion

Discuss any technique errors that occurred in the Exploration. Let students know that the writing samples may take much longer than the recently written pen tests.

Be sure students account for carefully choosing the best portions of the note for their analysis. Students should also suggest performing multiple trials on different small segments of the note. They should use parts where the writing is horizontal and there are no loops or vertical writing because these can complicate or obscure the chromatogram pattern.

Proposed Procedure

Use a pencil to sketch a circle about the size of a quarter in the center of the piece of filter paper. Write the numbers 1–4 in pencil around the inside of this circle, as shown in the illustration. Make a large dot with each pen on the circle beside that pen's number. Repeat with a second piece of filter paper.

Roll up two triangles of filter paper to be used as wicks. Use the pencil to poke a small hole in the center of one marked piece of filter paper. Insert a rolled-up wick through the hole. Fill the petri dish to the halfway point with water. Set the wick into this water and wait for the chromatogram to develop. Repeat with the second piece of filter paper and wick, but use isopropanol in the petri dish lid.

Allow each chromatogram to develop for approximately 15 min or until the solvent is about 1 cm from the outside edge of the paper. Remove each piece of filter paper from the petri dish and lid, and allow them to dry.

Make chromatograms with each solvent using different pieces of the ransom note.

Post-Lab

Disposal

Set out a disposal container for any isopropanol left over at the end of the procedure. Dilute it with 10 times as much water, and pour it down the drain. Students are instructed to pour the water down the drain. The chromatograms may be discarded in the trash can.

Analysis

Students' answers will vary depending on the pen you use to write the ransom note. Be certain students give a complete explanation of their evidence by comparing similar regions on the ransom note and the pen chromatograms.

Additional Notes

Testing for Dissolved Oxygen

11-2
Technique Builder

Objectives

Students will
- use appropriate lab safety procedures.
- use a Bunsen burner or a hot plate to heat water.
- measure temperature using a thermometer or a LEAP System thermistor probe.
- measure the concentration of oxygen in a sample of water using a dissolved oxygen meter, LEAP System probe, or test kit.
- graph the relationship between concentration and temperature.
- infer a general rule of thumb for gas solubilities and temperature.
- relate changes in gas solubility to a fish kill.

Planning

Recommended Time

2 lab periods (one day to prepare the water samples, and one for testing)

Materials

(for each lab group)
- Ice, about 200 g
- 600 mL beaker
- Beaker tongs
- Hot mitt
- Jars with screw-on lids, 4

Bunsen burner option
- Bunsen burner
- Gas tubing
- Ring stand
- Ring clamp
- Striker
- Wire gauze with ceramic center

Hot plate option
- Hot plate

LEAP System option
- LEAP System
- LEAP dissolved oxygen probe
- LEAP thermistor probe

Alternative option
- Dissolved oxygen test kit with 4 test ampuls, or dissolved oxygen meter
- Thermometer, nonmercury

Solution/Material Preparation

1. Canning jars (8 oz) are acceptable for use in holding the water samples. Be sure the

Required Precautions

- Goggles and a lab apron must be worn at all times to provide protection for the eyes and clothing.
- Tie back long hair and loose clothing.
- Read all safety cautions, and discuss them with your students.
- Remind students to use beaker tongs to pour the hot water because it can burn or scald.
- If the LEAP System with dissolved oxygen probe is used, the precautions listed in the teacher's notes on page T37 must be followed to avoid electric shock.

jars you use are heat resistant. If larger jars are used, larger amounts of water than those mentioned in the procedure will be necessary.

2. If your students will be using the LEAP System dissolved oxygen probe or a dissolved oxygen meter, be sure to refer to the information accompanying the equipment for detailed operating instructions. Be certain you know whether or not it is necessary to use a standard solution to calibrate the probe or meter so that you

can prepare the students for step **10** in the Technique section. If students share the probe or meter, as many as 15 lab groups can perform this Exploration with one probe or meter.

3. Do not use dissolved oxygen test kits that make use of the Winkler method or other titrimetric methods such as LaMotte's test kits because of the dangerous chemicals and the difficulties related to disposal. A simple and less expensive dissolved oxygen test kit is available from scientific suppliers such as Ward's Natural Science Establishment (1-800-962-2660). The kit contains small reagent-filled ampuls sealed in a vacuum. When the tip is broken off and dipped into a sample, the sample is drawn inside to react with the reagents, after which it can be compared to standards provided in the kit. When used properly, the user will not come into contact with any of the reagents. For this Exploration, each lab group needs four test ampuls. The catalog numbers and other information necessary for ordering these kits is as follows.

Ward's 21 W 9001 — 10 test ampuls and a set of standard ampuls

21 W 9002 — refill kit with 10 additional test ampuls

Student Orientation

Techniques to Demonstrate

Show students how to sample the water while disturbing it as little as possible, to avoid changing the O_2 concentration. Demonstrate the use of the dissolved oxygen test procedure you will use. If students will use the LEAP System probe or a dissolved oxygen meter, be sure to demonstrate how it should be calibrated (step **10** of the Technique section).

Sample Data	
Temp. (°C)	Dissolved O_2 (ppm)
4.0	10.1
24.4	8.3
50.2	5.3
100.1	1.1

If they will be using the dissolved oxygen test kits, show students how to pour 25 mL into the sample cup. Then, they can insert the ampul in one of the depressions in the bottom of the cup. Then, when they press the ampul against the side of the depression, the tip will snap off, and the ampul will fill with water. A small bubble will remain in the ampul. The ampul should be taken out of the cup with the open end downward. The open end should not be sharp, but to prevent injury, students should cover it with tape or a piece of tissue. They should mix the contents of the ampul by inverting it several times, allowing the bubble to travel from end to end. After two minutes, students can compare the color to those of the standards. Show how to use a white background behind the ampuls to compare the test colors to those of the standards.

Pre-Lab Discussion

Be certain students understand that solubility is not constant for all conditions, but changes when the temperature changes. Explain that to avoid contact with oxygen in the air as the samples change in temperature, it is important that they be kept in sealed containers that are almost entirely full and free of air.

Post-Lab

Disposal

The water samples may be rinsed down the drain. If you used dissolved oxygen test kits as described in the Solution/Material Preparation section, collect the opened, used ampuls for further treatment. Because the ampuls contain EDTA, they cannot be disposed in a landfill. Put the opened ampuls in an evaporating dish. In a hood known to be operating properly, heat them to at least 250°C for at least 15 minutes to decompose the EDTA. Ampuls that have been treated in this manner may be discarded in the waste basket. **Never heat sealed ampuls because they may burst open in a violent manner.**

Answers to

1.

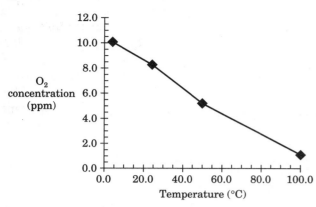

2. As the temperature increases, the solubility of oxygen decreases.

3. If the water was disturbed, it might make bubbles that could trap some oxygen.

4. As temperature increases, the amount of gas that the water can hold decreases. Any amount of gas over this amount is no longer soluble so it forms bubbles.

Conclusions

5. From the graph, the values appear to be about 9.0 ppm for 16°C and 7.9 ppm for 28 °C.

6. equation from the graphics calculator: O_2 conc. $= (-0.095)(temp.) + 10.4$

7. According to the equation, O_2 concentration is 8.9 ppm at 16°C and 7.7 ppm at 28°C.

8. $\dfrac{9.0 - 8.9}{9.0} \times 100 = 1\%$ error

$\dfrac{7.9 - 7.7}{7.9} \times 100 = 3\%$ error

9. graph answer: 9.0 ppm $\times \dfrac{90\%}{100\%} =$

8.1 ppm for damage

equation answer: 8.9 ppm $\times \dfrac{90\%}{100\%} =$

8.0 ppm for damage

By either method of evaluating the data, the warm water does not have enough oxygen.

Extensions

1. The attorney is incorrect because the solubility will follow the same general pattern for all samples of water, whether taken from the tap or not.

2. These tests do not conclusively establish that the fish kills were due only and solely to the water. They only establish that this warm water could be harmful to a fish.

3. Student answers may vary, but be sure that they are practical and will help isolate the cause of the fish kill. Students may suggest testing the water at the plant and in the lake and dissecting dead fish to discover the cause of death.

4. This will not work because the water will not be able to keep this additional oxygen dissolved at the warmer temperatures, and it will simply bubble away.

5. Student answers will vary, but they should indicate some way to restore the water to its original temperature or not to return it at all. Some students may suggest pouring the coolant water into containment ponds or tanks to cool off before being returned or recirculated.

Additional Notes

Solubility of Ammonia

EXPLORATION

11-3

Technique Builder

Objectives

Students will
- use appropriate lab safety procedures.
- use a Bunsen burner or a hot plate to heat water.
- measure temperature using a thermometer or a LEAP System thermistor probe.
- prepare a small amount of ammonia gas in a collection bulb.
- calculate amounts of ammonia present given a volume, temperature, and pressure.
- measure how long it takes for a quantity of ammonia to dissolve in water at different temperatures.
- relate changes in volume of a gas to the removal of its particles from the gas phase as they dissolve in water.
- compare the solubility of ammonia at two different temperatures.

Planning

Recommended Time

1 lab period

Materials

(for each lab group)
- $NH_3(aq)$, concentrated, 5 mL
- Distilled water
- Ice, 200 mL
- 250 mL beakers, 6
- 400 mL beaker
- Pipet, thin-stem
- Pipet, wide-stem
- Ruler, metric
- Scissors
- Stopwatch or clock with second hand
- Thermometer, nonmercury

Bunsen burner option
- Bunsen burner
- Gas tubing
- Ring stand
- Ring clamp
- Striker
- Wire gauze with ceramic center

Hot plate option
- Hot plate

LEAP System option
- LEAP System
- LEAP thermistor probe

Solution/Material Preparation

1. At the beginning of the lab, work in the hood, observe all required precautions, and pour about 100 mL of $NH_3(aq)$ into a

Required Precautions

- Wear safety goggles and a lab apron during the lab.
- Long hair and loose clothing must be tied back.
- Read all safety cautions, and discuss them with your students.
- All procedures of this lab must take place within a properly functioning fume hood because of the dangers involved with concentrated NH_3 and its fumes.
- Wear goggles and a lab apron when you pour the NH_3 into the 250 mL beaker. Work in a fume hood known to be in operating condition, with another person present nearby to call for help in case of an emergency.
- In case of an ammonia spill, dilute first with water, and be sure the room is well ventilated. Mop the spill up with wet cloths designated for spill cleanup while wearing disposable plastic gloves. Discard the cloths outside in a dumpster in the open air.
- If a hot plate or the LEAP System with thermistor probe is used, the precautions listed in the teacher's notes on page T37 must be followed to avoid electric shock.

single 250 mL beaker. Label the beaker conc. $NH_3(aq)$. This will be enough for the entire class to use. Be sure to keep the reagent bottle of concentrated $NH_3(aq)$ tightly sealed.

2. Check to be sure that your wide-stem pipets have bulbs with a volume of 3.5 mL. (The bulb will be between 44 and 48 mm long and about 13 or 14 mm in diameter). When the 4.0 cm of the stem is accounted for, total volume of the collecting bulb should be 5.0 mL. Most vendors will provide you with this information about their pipets. If your pipets are a different size or shape, you may have to calibrate them yourself. Calculate the total bulb volume by filling the bulb with water and then emptying it into a graduated cylinder.

3. Once a class set of collecting pipets have been used, they can be re-used again and again if they are emptied and shaken dry.

Student Orientation

Show students how to prepare the pipets to make collecting bulbs. With both pipets empty, demonstrate how the collecting bulb fits on top of the generating pipet. Show how to hold them so that air can escape from the collecting bulb. With an empty pipet bulb, demonstrate how to plunge the bulb neck-first into a beaker of water, deep enough so that the opening stays underwater in steps 11 and 12. Remind students to use distilled water for all of the water baths.

Pre-Lab Discussion

It may be necessary to review the discussion of gas solubility from pages 434–435. Stu-

Sample Data and Analysis

Students should come to the general conclusion that ammonia will dissolve better in cold water. Specific data may vary from the sample data shown here.

Temp. (°C)	Time (s)
0.0	1.0 s
25.0	2.0 s
50.0	2.7 s
80.0	3.5 s

dents often have the misconception that once the substance dissolves, they will see bubbles in the liquid. Remind them that a solution is a stable and homogeneous mixture. Point out that the fluid in a sealed soda bottle contains much dissolved carbon dioxide, but it bubbles only if the pressure is released by opening the bottle, making the CO_2 less soluble. The bubbles are due to the portion of the gas that is insoluble.

Post-Lab

Disposal

Allow the solutions to remain in the hood for several hours to allow most of the ammonia to vaporize. Working in the hood, combine them carefully, while stirring, into a single container. Adjust the pH of the liquid that remains with 1.0 M sulfuric or acetic acid until it is between 5 and 9. Then pour the liquid down the drain.

Answers to

Analysis and Interpretation

1. Because the same volume of gas contains the same amount of particles if it is at the same temperature and pressure, density will be directly related to molar mass. O_2 and N_2 are much denser than NH_3 and will be displaced from the bulb.

2. As the concentrated solution was heated, the solubility of NH_3 decreased. The excess NH_3 that was no longer soluble bubbled out of solution.

3. volume = 0.0050 L
temperature = $273 + 25°C = 298$ K
pressure = 1.0 atm
$R = 0.0821$ L·atm/mol·K
$PV = nRT$

$$n = \frac{PV}{RT} =$$

$$\frac{(1.0 \text{ atm})(0.0050 \text{ L})}{(0.0821 \text{ L·atm/mol·K})(298 \text{ K})} =$$

2.0×10^{-4} mol NH_3

4. 2.0×10^{-4} mol $NH_3 \times \dfrac{17.04 \text{ g } NH_3}{1 \text{ mol}} =$

3.4×10^{-3} g NH_3

5. When the NH_3 came into contact with the water, some of it dissolved. With

fewer NH_3 particles, the pressure was decreased, so the water was pulled into the bulb. As all of the NH_3 dissolved, the water filled the bulb.

Conclusions

6. The best temperature would be 0.0°C because the NH_3 dissolves much more rapidly at that temperature.

7. Using the sample data, it should take about half as long, or 7.5 min, to dissolve 500. g of ammonia gas at 0.0°C.

8. The data do not measure the total solubility of NH_3, but only how quickly a given quantity dissolved. In fact, this quantity of NH_3 was so small that it dissolved completely in all of the samples of water, as indicated by the way that water filled all of the bulbs completely.

9. NH_3 is much more soluble in water than CO_2 is because NH_3 is a polar molecule like water.

Extensions

1. At 25°C, the partial pressure of water vapor is 3.17 kPa.

   101.325 kPa − 3.17 kPa = 98.2 kPa

   $$\frac{98.2 \text{ kPa}}{101.325 \text{ kPa/atm}} = 0.969 \text{ atm}$$

$$n = \frac{PV}{RT} =$$

$$\frac{(0.969 \text{ atm})(0.0050 \text{ L})}{(0.0821 \text{ L} \cdot \text{atm/mol} \cdot \text{K})(298 \text{ K})} =$$

2.0×10^{-4} mol NH_3

Rounded to the correct number of significant figures, there is no measurable difference in the mole amount.

2. To maximize the efficiency of the dissolving process, they should perform it at the highest possible pressure. At 2.0 atm, twice as much NH_3 should dissolve.

3. Students' suggestions for improving the procedure will vary. Possibilities include running multiple trials. Be sure answers are safe and include carefully planned procedures.

4. If average atmospheric temperatures are raised and this change is carried through to ocean waters, less CO_2 would dissolve in the water, and there would be more of it released to the atmosphere.

Additional Notes

Solubility Product Constant

EXPLORATION

12-1
Technique Builder

Objectives

Students will
- prepare a saturated solution of sodium chloride.
- determine the mass of sodium chloride dissolved in a volume of saturated solution.
- calculate the solubility of sodium chloride.
- determine the solubility product constant of sodium chloride.
- observe the effect of adding a common ion to a saturated solution.

Planning

Recommended Time

1 lab period

Materials

(for each lab group)
- Distilled water
- NaCl, 15 g
- $Na_3C_6H_5O_7 \cdot 2H_2O$, 0.50 g
- 150 mL beakers, 2
- 10 mL graduated cylinder
- 25 × 100 mm test tube
- Balance
- Crucible tongs
- Evaporating dish
- Glass stirring rod
- Spatula
- Weighing paper, 2 pieces

Bunsen burner option
- Bunsen burner
- Gas tubing
- Ring stand and ring
- Striker
- Wire gauze with ceramic center, 2

Hot plate option
- Hot plate

Solutions/Materials Preparation

1. Sodium citrate is actually used as a food additive in some products. However, if it is not in stock, the common-ion portion of the lab (steps **10–11**) can be performed with the following other sodium salts instead of sodium citrate: sodium carbonate, sodium hydrogen carbonate, sodium bromide, sodium iodide, sodium oxalate, sodium sulfate, sodium thiosulfate, and

Required Precautions

- Goggles and a lab apron must be worn at all times.
- Tie back long hair and loose clothing when working in the lab.
- Read all safety cautions, and discuss them with your students.
- If the hot plate is used, the precautions described in the teacher's notes on page T37 must be followed to avoid electric shock.

sodium acetate. Tell the students which one will be used so that they can adjust their calculations accordingly. **Do not use sodium nitrate or other sodium salts besides those listed because of the difficulties involved in disposal.**

2. Instead of decanting in step **3,** you may have students use micropipets to transfer an appropriate amount of the solution to the next beaker for testing. Be certain students do not place the tip of the pipet near the solid residue, or some of the solid will be drawn into the pipet and the results will not be as accurate.

Student Orientation

Techniques to Demonstrate

Have the equipment set up for heating the evaporating dish. If you are using a Bunsen burner, show students the appropriate size and type of flame needed to heat the salt solution slowly and avoid spattering. If you are using a hot plate, suggest a setting that will heat slowly without spattering.

Demonstrate the decanting step, and explain how important it is to avoid getting crystals of NaCl in the saturated solution that will be tested. If you will be using micropipets instead of decanting, point out how important it is to keep the tip of the pipet well above the solid so that none is accidentally included.

Be certain students understand that when they decant the saturated solution in step **3**, it is the solution they want to keep, not any solid that remains undissolved. Students sometimes become confused about what they are supposed to do in this step.

Discuss the nature of solubility equilibrium in terms of the undissolved solid on the bottom of a beaker dissolving into the solution at the same rate that the dissolved solute precipitates out of solution. Students must understand that the apparently static nature of the system on the macroscopic level does not reflect the dynamic equilibrium at the microscopic level.

Post-Lab

Disposal

Set out one disposal container for students. It should contain only sodium chloride and another sodium salt. If the other salt is one recommended in the Solution/Materials Preparation section, the salts can be dissolved in water and poured down the drain.

Answers to

Analysis and Interpretation

1. $44.50 \text{ g} - 41.36 \text{ g} = 3.14 \text{ g NaCl}$

2. $3.14 \text{ g} \times \dfrac{1 \text{ mol NaCl}}{58.44 \text{ g}} =$

   $\qquad\qquad\qquad\qquad 0.0537 \text{ mol NaCl}$

Sample Data	
Mass of empty evaporating dish	41.36 g
Volume of saturated NaCl solution	10.0 mL
Mass of evaporating dish and NaCl	44.50 g
Mass of $Na_3C_6H_5O_7 \cdot 2H_2O$	0.49 g

3. $\dfrac{0.0537 \text{ mol NaCl}}{10.0 \text{ mL}} \times \dfrac{1000 \text{ mL}}{1 \text{ L}} =$

   $\qquad\qquad\qquad\qquad\qquad 5.37 \text{ M}$

4. The equilibrium equation for dissolving NaCl to form a saturated solution is $NaCl(s) \rightleftharpoons Na^+(aq) + Cl^-(aq)$.

5. The equation in item **4** shows a 1:1 ratio between NaCl and Na^+ and between NaCl and Cl^-. NaCl is a strong electrolyte, which dissolves completely. Therefore, $[Na^+] = [Cl^-] = 5.37$ M.

6. $K_{sp} = [Na^+][Cl^-]$

7. $K_{sp} = (5.37)(5.37) = 28.8$

8. $0.49 \text{ g } Na_3C_6H_5O_7 \cdot 2H_2O \times \dfrac{1 \text{ mol}}{294.12 \text{ g}} =$

   $\qquad\qquad\qquad\qquad 1.7 \times 10^{-3} \text{ mol}$

Conclusions

9. $4000.0 \text{ L} \times \dfrac{5.37 \text{ mol NaCl}}{1 \text{ L}} \times$

   $\dfrac{58.44 \text{ g}}{1 \text{ mol}} = 1.26 \times 10^6 \text{ g (or 1260 kg)}$

10. The addition of $Na_3C_6H_5O_7$ to a saturated solution of NaCl increases the concentration of Na^+. Any increase in concentration of either ion causes the numerical value of K_{sp} to be exceeded. Therefore, NaCl precipitates from the solution until the product of the concentrations of the ions again equals the K_{sp}.

11. The additives that will exhibit the common-ion effect with a NaCl solution are the ones most likely to have an effect. These are calcium chloride, sodium bicarbonate, and sodium nitrate.

Extensions

1. If the NaCl brine is diluted to half of its concentration, $[Na^+] = [Cl^-] = 2.68$ M.
   $\dfrac{50.0 \text{ g } CaCl_2}{1 \text{ L}} \times \dfrac{1 \text{ mol } CaCl_2}{110.98 \text{ g}} = 0.451 \text{ M}$
   $[Cl^-]$ from $CaCl_2 = 0.902$ M
   total $[Cl^-] = 2.68$ M $+ 0.902$ M $= 3.58$ M
   Using the equilibrium expression,
   $[Na^+][Cl^-] = (2.68)(3.58) = 9.59$.
   Because K_{sp} (28.8 at this temperature) is not exceeded, no precipitation will occur.

2. The large solubility product constants of

iron(III) nitrate and iron(III) acetate mean that they are very soluble. These compounds are too quickly removed by rain and irrigation to be useful for boosting the iron content of soil.

3. $[Na^+]$ from $Na_3C_6H_5O_7 =$

$$3 \times \frac{1.7 \times 10^{-3}\,mol}{0.005\,L} = 1.02\,M$$

total possible $[Na^+] = 5.37\,M + 1.02\,M = 6.39\,M$

Assuming that x mol/L of Na^+ and Cl^- ions will precipitate as solid NaCl, the equilibrium expression has the following values.

$K_{sp} = 28.8 = [Na^+][Cl^-] =$
$$(6.39 - x)(5.37 - x)$$

$0 = x^2 - 11.76x + 5.5$

$$x = \frac{11.76 \pm \sqrt{(11.76)^2 - 4(5.5)}}{2}$$

$x = 0.5$ or 11.3 (11.3 is impossible because that much was not originally available)

$[Na^+] = 5.89\,M$

$[Cl^-] = 4.87\,M$

4. Student answers will vary, but students should see that the equilibrium equation that follows is a key part of the process.

$CaCO_3(s) + H_2O(aq) + CO_2(g) \rightleftharpoons$
$Ca^{2+}(aq) + 2HCO_3^-(aq)$

When calcium carbonate is exposed to air, a small amount of it reacts with and dissolves in water due to CO_2 in the air, forming the aqueous solution indicated on the right side of the equation. After the water seeps into a cave, which typically has a lower amount of CO_2 in its air, the reverse reaction occurs, leaving a layer of insoluble $CaCO_3$ behind.

5. Although the solubility limits of sucrose can be determined experimentally, sucrose does not dissociate. Thus, a solubility product constant cannot be determined and there should be no common-ion effect.

6. The solubility of NaCl changes very little with temperature, so a temperature control device should be unnecessary. K_{sp} should not change much.

7. $pK_{sp} = -\log K_{sp}$
$-\log (2.5) = -0.40$
$-\log (25.7) = -1.41$
$-\log (132) = -2.12$

Solubility Product Constant—Algae Blooms

INVESTIGATION 12-1

Problem Solving

Objectives

Students will
- use appropriate lab safety procedures.
- prepare saturated solutions of $CuSO_4$ and $CuCl_2$.
- demonstrate proficiency in decanting and other methods of handling liquids.
- measure volume using a graduated cylinder.
- measure mass using a laboratory balance.
- relate concentration of each saturated solution to the mass of residue remaining after evaporation.
- calculate the solubility product constant, K_{sp}, of the two salts.
- calculate how many liters of saturated solution would be necessary to make an entire 3.9×10^6 L pond a 0.05 M solution of Cu^{2+} ions.
- calculate mass of salt needed to treat a pond.
- calculate costs related to these amounts of salt.
- evaluate which salt would be most effective.

Planning

Recommended Time
1–2 lab periods (depends upon availability of two hot plates per lab group)

Prerequisite
Exploration 12-1

Materials
(for each lab group)
- $CuCl_2 \cdot 2H_2O$, 25 g
- $CuSO_4 \cdot 5H_2O$, 15 g
- Distilled water
- 150 mL beakers, 4
- 10 mL graduated cylinder
- Balance
- Crucible tongs
- Evaporating dishes, 2
- Glass stirring rods, 2
- Spatula

Bunsen burner option
- Bunsen burner
- Gas tubing
- Ring stand and ring
- Striker
- Wire gauze with ceramic center

Required Precautions
- Goggles and a lab apron must be worn at all times.
- Tie back long hair and loose clothing.
- Read all safety cautions, and discuss them with your students.
- Remind students that heated objects can be hot enough to burn even if they look cool. Students should always use crucible tongs when handling any lab equipment that has been heated.
- If a hot plate is used, the precautions listed on page T37 must be followed to avoid electric shock.

Hot plate option
- Hot plate

Estimated cost of materials:
$117 000–$132 000

	$CuSO_4$	$CuCl_2$
Mass of evaporating dish	41.36 g	41.36 g
mL of saturated solution	10.0 mL	10.0 mL
Mass of evaporating dish and salt	43.21 g	47.06 g

For both salts

$$3.90 \times 10^6 \text{ L} \times \frac{0.0500 \text{ mol}}{1 \text{ L}} =$$

$$1.95 \times 10^5 \text{ mol of salt needed for a pond}$$

For $CuSO_4$

Molarity of solution:

$$\frac{1.85 \text{ g}}{10.0 \text{ mL}} \times \frac{1 \text{ mol } CuSO_4}{159.62 \text{ g}} \times \frac{1000 \text{ mL}}{1 \text{ L}} =$$

$$1.16 \text{ M}$$

$K_{sp} = [Cu^{2+}][SO_4{}^{2-}] = (1.16)^2 = 1.35$

Volume of saturated solution needed:

$$(1.95 \times 10^5 \text{ mol}) \times \frac{1 \text{ L}}{1.16 \text{ mol}} = 1.68 \times 10^5 \text{ L}$$

Mass of hydrated salt needed:

$$(1.95 \times 10^5 \text{ mol}) \times \frac{249.72 \text{ g}}{1 \text{ mol}} \times \frac{1 \text{ kg}}{1000 \text{ g}} =$$

$$4.87 \times 10^4 \text{ kg}$$

Cost: $(4.87 \times 10^7 \text{ g}) \times \dfrac{\$100}{1 \text{ g}} = \$4.87 \times 10^9$

For $CuCl_2$

Molarity of solution:

$$\frac{5.70 \text{ g}}{10.0 \text{ mL}} \times \frac{1 \text{ mol } CuCl_2}{134.45 \text{ g}} \times \frac{1000 \text{ mL}}{1 \text{ L}} =$$

$$4.24 \text{ M}$$

$K_{sp} = [Cu^{2+}][Cl^-]^2 = (4.24)(8.48)^2 = 305$

Volume of saturated solution needed:

$$(1.95 \times 10^5 \text{ mol}) \times \frac{1 \text{ L}}{4.24 \text{ mol}} = 4.60 \times 10^4 \text{ L}$$

Mass of hydrated salt needed:

$$(1.95 \times 10^5 \text{ mol}) \frac{170.49 \text{ g}}{1 \text{ mol}} \times \frac{1 \text{ kg}}{1000 \text{ g}} =$$

$$3.32 \times 10^4 \text{ kg}$$

Cost: $(3.32 \times 10^7 \text{ g}) \times \dfrac{\$200}{1 \text{ g}} = \$6.64 \times 10^9$

Students may chose $CuCl_2 \cdot 2H_2O$ as the best choice to kill algae because less mass and smaller volumes of saturated solution are needed. Alternatively, $CuSO_4 \cdot 5H_2O$ costs less per gram. A mixture will not provide a greater concentration because of the common-ion effect.

Solution/Material Preparation

The lab can be performed with one hot plate per lab group, but it is more likely that students will finish in one lab period if two hot plates are available.

Student Orientation

Techniques to Demonstrate

Show students how to gently heat the saturated solutions to avoid spattering. When the solutions appear dry, students should continue to heat them gently until $CuSO_4$ is white and $CuCl_2$ is light brown. If a yellow color persists in the $CuSO_4$ after it has cooled, that is a sign that the heating was too rapid and decomposition occurred.

Pre-Lab Discussion

Begin by discussing the results and procedure used in the Exploration, especially any errors in lab technique that occurred. Make sure they understand the answers to the Conclusions for the Exploration.

Some students may decide they don't need to do this lab at all because the K_{sp} values are available in reference books. Tell them that students need data to support their conclusions because the K_{sp} values from reference sources are obtained under ideal conditions with very pure reagents.

Tips for Evaluating the Pre-Lab Requirements

Be certain that students have developed a procedure that details exactly how they will divide the lab work. For instance, one member of the group can be preparing one solution and evaporating it while another member prepares the other solution. Otherwise, this lab will be difficult to finish in time.

Proposed Procedure

Place 25 mL of distilled water in a 150 mL beaker. Add 15 g of $CuCl_2 \cdot 2H_2O$, and stir until it dissolves. Continue to add salt in 5 g increments, while stirring, until some remains undissolved on the bottom. Repeat the procedure for $CuSO_4 \cdot 5H_2O$ in another 150 mL beaker.

Decant the saturated solutions into two other labeled beakers.

Measure and record the masses of two labeled evaporating dishes.

Measure 10.0 mL of each solution into separate evaporating dishes, and record the volumes of the solutions, washing the graduated cylinder between uses.

Using a Bunsen burner or hot plate, gently evaporate the water from each solution in the evaporating dishes. Continue heating until the salts are dry and anhydrous.

Allow the evaporating dishes to cool and obtain the masses of the evaporating dishes and salts.

Post-Lab

Disposal

Provide two labeled containers for the disposal of $CuSO_4$ and $CuCl_2$ solutions and any excess of the original compounds. Later, redissolve the contents of the container in distilled water. Let the solution evaporate until dry, and then recover the crystals for re-use next year. Do not dispose of these compounds in a landfill or incinerator or down the drain.

Additional Notes

Freezing-Point Depression—Testing De-Icing Chemicals

EXPLORATION

12-2

Technique Builder

Objectives

Students will
- use appropriate lab safety procedures.
- measure temperature using a thermometer or the LEAP System with thermistor probe.
- determine the freezing-point depression of three solutions.
- calculate the molality of the particles of the solutions.
- evaluate the effectiveness of each solute as a de-icer based on freezing-point depression per gram and per dollar.
- determine the molar mass of each solute given the number of particles in a formula unit.

Planning

Recommended Time

1 lab period

Materials

(for each lab group)
- De-icer F-38 (frozen solution of $CaCl_2$)
- De-icer H-22 (frozen solution of $NaCl$)
- De-icer J-27 (frozen solution of sucrose)
- Ice (frozen distilled water)
- 250 mL beaker
- 25 mL graduated cylinder
- Glass stirring rod
- Test tubes (large), 3
- Test-tube rack
- Thermometer, nonmercury

LEAP System option
- LEAP System with thermistor probe

Solution/Material Preparation

1. Students will also need equipment for crushing the frozen solutions. Possibilities include toweling or a plastic bag and hammer.
2. Prepare the solutions the day before the lab, and allow them to freeze overnight. Rinse out two empty 1 or 2 qt cardboard milk cartons. Cut each in half lengthwise to make four "trays." For each solution, dissolve 100.0 g in exactly 1.00 L of water. Fill each tray about halfway with a solution or distilled water. **Put the trays**

Required Precautions

- Goggles and a lab apron must be worn at all times.
- Read all safety cautions, and discuss them with your students.
- If the LEAP System with thermistor probe is used, the precautions listed in the teacher's notes on page T37 must be followed to avoid electric shock.

in the freezer compartment of a refrigerator DESIGNATED TO BE USED ONLY FOR CHEMICAL STORAGE. Do NOT use a refrigerator in which food was, is, or will be stored.

Student Orientation

Techniques to Demonstrate

Show students how you want them to crush their samples of solid solutions. This could be by wrapping it in toweling and smashing it on the floor or by placing the cube of solid into a clean plastic sandwich bag and tapping it *gently* with a hammer.

Pre-Lab Discussion

Review the definition of molality, and stress how it differs from molarity. Emphasize the idea that for colligative properties, it is the ratio of particles of solute to particles of solvent that is important.

Be certain students understand that a substance's melting point and freezing point are the same.

Post-Lab

Disposal

Set out three disposal containers, one for each of the solutions (F-38, H-22, and J-27). The solutions may be refrozen and reused if they have not been contaminated. The water can be evaporated and the salts reused. For disposal, these substances can be dissolved and poured down the drain.

Answers to

Analysis and Interpretation

1. $\Delta t_{F\text{-}38} = -5.0°C$
$\Delta t_{H\text{-}22} = -6.6°C$
$\Delta t_{J\text{-}27} = -0.5°C$

2. $m \times n = \dfrac{\Delta t_f}{K_f}$

for F-38: $\dfrac{-5.0°C}{-1.86°C/m} = 2.7\ m$

for H-22: $\dfrac{-6.6°C}{-1.86°C/m} = 3.6\ m$

for J-27: $\dfrac{-0.5°C}{-1.86°C/m} = 0.27\ m$

3. The melting/freezing point is the temperature at which the solid and liquid phases are in equilibrium.

4. Freezing-point depression and other concentration-dependent properties of solutions are intensive properties for a given solution, no matter how large the sample.

Conclusions

5. Using the amount of freezing-point depression per gram of solute as the selection criterion, H-22 is the best choice.

Sample Data and Analysis

Substance	Temp. (°C)
H_2O	0.1
F-38	−4.9
H-22	−6.5
J-27	−0.4

6. H-22 is the cheapest solute by mass.

7. For a most accurate reading, the thermometer must be surrounded on all sides by the solid-liquid mixture.

8. The most important reason for determining the freezing point of water is to calibrate the thermometer.

Extensions

1. F-38: $\dfrac{3\text{ particles}}{2.7\ m} \times \dfrac{100.0\text{ g}}{1\text{ kg solvent}} =$

110 g/mol

H-22: $\dfrac{2\text{ particles}}{3.6\ m} \times \dfrac{100.0\text{ g}}{1\text{ kg solvent}} =$

56 g/mol

J-27: $\dfrac{1\text{ particle}}{0.27\ m} \times \dfrac{100.0\text{ g}}{1\text{ kg solvent}} =$

370 g/mol

2. The 1.0 m $CuCl_2$ solution should have a greater freezing-point depression because it will form three particles per formula unit instead of two.

3. 2.0 mol sucrose $\times \dfrac{1\text{ kg }H_2O}{1\text{ mol sucrose}} \times$

$\dfrac{1000\text{ g}}{1\text{ kg}} \times \dfrac{1\text{ mol}}{18.02\text{ g}} \times \dfrac{6.02 \times 10^{23}}{1\text{ mol}} =$

6.7×10^{25} molecules

4. Students' suggestions for improving the procedure will vary. Students may propose that liquid solutions of the same molality could be added to the solid solutions to create the solid-liquid freezing mixture. Be sure answers are safe and include carefully planned procedures.

5. Students' answers will vary, but students should notice that salt and other de-icing agents can damage pavement, plants, and automobiles.

6. $-30°C \times \dfrac{1\ m}{-1.86°C} = 16\ m$

7. In hot weather, antifreeze containing ethylene glycol raises the boiling point of the engine coolant and makes it less likely that the engine will overheat due to the engine coolant boiling over.

8. $5.50°C - \left(0.750\ m \times \dfrac{5.12°C}{1\ m}\right) =$

$5.50°C - 3.84°C = 1.76°C$

Freezing-Point Depression—Making Ice Cream

INVESTIGATION 12-2

Problem Solving

Objectives

Students will
- use appropriate lab safety procedures.
- make saturated solutions of three different compounds.
- measure freezing-point depression for each of the solutions.
- measure temperatures using a thermometer or LEAP System with thermistor probe.
- measure mass using a lab balance.
- measure volume using a graduated cylinder.
- determine molality and molarity from experimental measurements.
- calculate the value of the solubility product constant.
- calculate the theoretical value of the freezing-point depression.
- calculate the percentage difference between theoretical and actual values for freezing-point depression.
- design and implement their own procedure.

Planning

Recommended Time
1 lab period, if students divide the work within their lab group

Prerequisite
Exploration 12-1 and Exploration 12-2

Materials
(for each lab group)
- $Ca(C_2H_3O_2)_2 \cdot 2H_2O$, 20.0 g
- Distilled water
- NaCl, 20.0 g
- $NaHCO_3$, 5.0 g
- Ice, 500 mL
- 250 mL beakers, 3
- 100 mL graduated cylinder
- Balance
- Beaker tongs
- Crucible tongs
- Evaporating dishes, 3
- Glass stirring rods, 3
- Thermometer, nonmercury

LEAP System option
- LEAP System
- LEAP Thermistor probe

Bunsen burner option
- Bunsen burner

Required Precautions

- Always wear goggles and a lab apron to provide protection for eyes and clothing.
- Read all safety cautions, and discuss them with your students.
- Confine long hair and loose clothing.
- Remind students that when equipment has been heated, it should be handled with tongs or a hot mitt. Only crucible tongs should be used with an evaporating dish, and only beaker tongs should be used with a beaker.
- If a hot plate or the LEAP system with thermistor probe is used the precautions listed on page T37 must be followed to avoid electric shock.

- Gas tubing
- Ring stand and ring
- Striker
- Wire gauze with ceramic center

continued on next page

Hot plate option
• Hot plate
Estimated cost of materials:
$160 000–$170 000
Student invoices should be $15 000 more than their estimated costs. It is acceptable to include labor costs in the invoice.

Solution/Material Preparation

The procedure works equally well with a plastic-foam cup instead of a beaker.

Student Orientation

Techniques to Demonstrate

You may need to remind students that they will need to measure the masses of all evaporating dishes before solution is added. Then, they will need to measure the mass of the solution before the solvent is evaporated, as well as after. They should also measure the initial volume of the solution. Many students will forget that they must measure

Sample Data and Analysis

Substance	NaCl	$Ca(C_2H_3O_2)_2$	$NaHCO_3$
Temperature (°C)	− 15.0	− 9.9	− 2.3
Mass of empty evaporating dish (g)	42.00	41.55	41.87
Volume of solution (mL)	50.0	49.5	51.3
Mass of dish + solution (g)	99.40	111.39	95.22
Mass of dish + crystals (g)	53.48	65.55	45.07

Temperature of pure ice water: 0.0°C

For NaCl:

$$11.48 \text{ g NaCl} \times \frac{1 \text{ mol NaCl}}{58.44 \text{ g}} = 0.1964 \text{ mol NaCl}$$

$$\frac{0.1964 \text{ mol Na}^+}{0.500 \text{ L}} = 3.93 \text{ M Na}^+ = 3.93 \text{ M Cl}^-; K_{sp} = 15.4$$

$$\Delta t_{expected} = \frac{0.1964 \text{ mol NaCl}}{0.04592 \text{ kg H}_2\text{O}} \times \frac{2 \text{ particles}}{1 \text{ mol NaCl}} \times -1.86°\text{C}/m = -15.9°\text{C}$$

$$\% \text{ difference} = \frac{-15.9°\text{C} - (-15.0°\text{C})}{-15.0°\text{C}} \times 100 = 6\%$$

For $Ca(C_2H_3O_2)_2$:

$$14.00 \text{ g Ca(C}_2\text{H}_3\text{O}_2)_2 \times \frac{1 \text{ mol Ca(C}_2\text{H}_3\text{O}_2)_2}{158.18 \text{ g}} = 0.088\,51 \text{ mol Ca(C}_2\text{H}_3\text{O}_2)_2$$

$$\frac{0.088\,51 \text{ mol Ca}^{2+}}{0.0495 \text{ L}} = 1.79 \text{ M Ca}^{2+}; 3.58 \text{ M C}_2\text{H}_3\text{O}_2{}^-; K_{sp} = 6.41$$

$$\Delta t_{expected} = \frac{0.088\,51 \text{ mol Ca(C}_2\text{H}_3\text{O}_2)_2}{0.04584 \text{ kg H}_2\text{O}} \times \frac{3 \text{ particles}}{1 \text{ mol Ca(C}_2\text{H}_3\text{O}_2)_2} \times -1.86°\text{C}/m = -10.8°\text{C}$$

$$\% \text{ difference} = \frac{-10.8°\text{C} - (-9.9°\text{C})}{-9.9°\text{C}} \times 100 = 9\%$$

For $NaHCO_3$:

$$3.20 \text{ g NaHCO}_3 \times \frac{1 \text{ mol NaHCO}_3}{84.01 \text{ g}} = 0.0381 \text{ mol NaHCO}_3$$

$$\frac{0.0381 \text{ mol Na}^+}{0.0513 \text{ L}} = 0.743 \text{ M Na}^+ = 0.743 \text{ M HCO}_3{}^-; K_{sp} = 0.552$$

$$\Delta t_{expected} = \frac{0.0381 \text{ mol NaHCO}_3}{0.05015 \text{ kg H}_2\text{O}} \times \frac{2 \text{ particles}}{1 \text{ mol NaHCO}_3} \times -1.86°\text{C}/m = -2.82°\text{C}$$

$$\% \text{ difference} = \frac{-2.82°\text{C} - (-2.3°\text{C})}{-2.3°\text{C}} \times 100 = 23\%$$

the melting/freezing point of ice and distilled water.

Pre-Lab Discussion

Begin by discussing the results and procedures used in the Explorations, especially any errors in lab technique that occurred.

Review colligative properties, making certain that students understand that when ionic substances dissociate in water, more than one mole of particles can be formed from one mole of formula units, and all of these particles must be accounted for in calculating freezing-point depressions.

$\Delta t_{fp} = m \times n \times K_f$, where n is the number of particles in one formula unit

Review the differences between molarity and molality and how to calculate each one.

If students remain uncertain about how to proceed, you may provide the following leading questions for them to consider as they make their plans.

- How can you be sure a solution is saturated?
- What quantities must you measure to determine molarity?
- What quantities must you measure to determine molality?

Tips for Evaluating the Pre-Lab Requirements

Be certain students have clearly indicated in their data tables that they will measure the quantities indicated in the sample data table, especially the mass of the empty evap-

orating dishes. In order to make 50 mL of saturated solution, students should need slightly less than 20 g each of NaCl and $Ca(C_2H_3O_2)_2$ and 5 g of $NaHCO_3$. If students' plans suggest the use of more than these amounts, ask them to reconsider their plans.

Proposed Procedure

Measure the temperature of a mixture of ice and distilled water. Prepare a saturated solution by adding 20.0 g NaCl to a beaker containing ice and about 50 mL of water. Be certain that there is some NaCl left undissolved on the bottom and some ice floating on top. Measure and record the temperature of the solution. Measure and record the mass of an empty evaporating dish. Carefully decant some of the solution into a graduated cylinder. Record the volume. Pour the solution in the graduated cylinder into an evaporating dish. Measure and record the mass of the evaporating dish and solution. Gently heat the evaporating dish until only dry crystals remain. Allow to cool, then measure and record the mass of the evaporating dish and crystals. Repeat the procedure for $Ca(C_2H_3O_2)_2$ and $NaHCO_3$.

Post-Lab

Disposal

All of the solutions used in this lab can be washed down the drain with an excess of water.

Additional Notes

Measuring pH—Home Test Kit

INVESTIGATION 13-1

Problem Solving

Objectives

Students will
- use appropriate lab safety procedures.
- prepare extracts of natural products to use as potential pH indicators.
- add measured quantities of acid or base to adjust pH of a solution.
- measure pH using pH paper, a pH meter, or LEAP System with pH probe.
- evaluate the color changes of several potential pH indicators.
- develop a pH home test kit to measure pH over a wide range.
- design and implement their own procedure.

Planning

Recommended Time
2 lab periods (one for preparing potential indicators, one for testing them)

Prerequisite
None

Materials
(for each lab group)*
- 1.0 M NaOH, 50 mL
- 1.0 M HCl, 50 mL
- Distilled water
- 400 mL beakers, 6
- 250 mL beakers or Erlenmeyer flasks, 6
- 100 mL graduated cylinder
- Beaker tongs
- Buret clamp (double)
- Buret tubes, 2
- Glass stirring rod
- pH paper, narrow range, 30 small pieces
- Ring stand and ring
- Thermometer, nonmercury
- Wash bottle

pH meter option
- pH meter

LEAP System option
- LEAP System
- LEAP pH probe
- LEAP Thermistor probe

Bunsen burner option
- Bunsen burner
- Gas tubing
- Striker
- Wire gauze with ceramic center

Required Precautions
- Always wear goggles and a lab apron to provide protection for eyes and clothing.
- Long hair and loose clothing must be tied back.
- Read all safety cautions, and discuss them with your students.
- Students should not handle concentrated acid or base solutions.
- Wear goggles, a face shield, impermeable gloves, and a lab apron when you prepare the HCl and NaOH solutions. For the HCl solution, work in a hood known to be in operating condition. For both solutions, work with another person present nearby to call for help in case of an emergency. Be sure you are within 30 s walking distance of a properly working safety shower and eyewash station.

continued on next page

Hot plate option
- Hot plate

Estimated cost of materials:
$104 000–$124 000

* Note: if each lab group tests fewer than six substances, the amount of glassware and the cost of materials will be less.

Required Precautions

(continued)

- In case of an acid or base spill, first dilute with water. Then mop up the spill with wet cloths or a wet cloth mop while wearing disposable plastic gloves. Designate separate cloths or mops for acid and base spills.
- Remind students that the beakers used to prepare the indicators will be hot. They should use beaker tongs when moving these beakers.
- If a hot plate, pH meter, or the LEAP System with pH and thermistor probe are used, the precautions listed in the teacher's notes on page T37 must be followed to avoid electric shock.

1. To prepare 1.00 L of 1.0 M NaOH, observe the required precautions. Slowly add 4.0 g of NaOH to about 400 mL of distilled water and stop to stir it in order to avoid overheating. Add more distilled water to bring the total volume up to 1.00 L of solution.

2. To prepare 1.00 L of 1.0 M HCl, observe the required precautions. Slowly add 8.6 mL of concentrated HCl to approximately 400 mL of distilled water and stop to stir it in order to avoid overheating. Add more distilled water to bring the volume up to 1.00 L of solution.

3. Encourage students to choose a variety of items not on the reference list to test. Other possibilities include red apple skin, blueberries, red cherries, red onion, yellow onion, peach skin, pear skin, radish skin, rhubarb skin, tomato skin, or turnip skin.

4. Before lab, chop the samples of food products in a food processor that has been

Sample Data and Analysis

pH	Apple skin (red)	Beets	Blueberries	Cabbage (red)	Cherries	Onions (red)	Radish skin
2	orange	red	red	red	red	pink	orange
3				pink	orange		
4	pink		purple	lt. purple		colorless	pink
5			green	blue	brown		rose
6	yellow	purple				yellow	purple
7					green		
8				gray/green			dark purple
9							
10							brown
11		brown		lt. green			
12							

Depending on student choices for materials to test, the pH values will differ. Red cabbage and radish skin are the best universal indicators shown here. (Rhubarb skin and turnip skin are also good indicators with many different color changes.) However, students may conclude that the vegetable that is cheapest or easiest to obtain is the best indicator. Be sure student plans for manufacturing the indicators are practical and straightforward.

dedicated for use ONLY in the lab for nonfood products. Otherwise, a different method of chopping must be used. Each 200 mL of indicator will require approximately 250 mL of chopped sample.

5. To save time, you can assign potential indicators. Each lab group can scale up the instructions to prepare 1.5 L of a specific indicators. On the second day, each lab group can test 100 mL samples of all of the indicators.

6. This investigation can be performed as a microscale procedure. Instead of beakers, prepare several spot plates containing solutions with pH values from 2 to 12. Then, using a pipet, add a drop of the indicator to be tested to each of the spot plates. Instead of each student preparing six indicators, the teacher can prepare six or more. One beaker's worth should be enough for the whole class.

Student Orientation

Techniques to Demonstrate

Show students how to prepare the indicator samples as described in the memorandum.

Be certain students realize that they can vary the pH by adding HCl or NaOH very slowly.

Demonstrate the calibration of the pH meter or the LEAP pH probe.

Pre-Lab Discussion

Students should be advised to allow the samples to cool before testing and to test a consistent amount of each solution.

If students are uncertain about how to proceed, you may provide the following leading questions for them to consider as they plan their procedure.

- How can you use the NaOH and HCl to adjust the pH so that you will be able to see all of the possible color changes?
- Which would be easier to adjust to a pH of exactly 2.0 using 1.0 M HCl: a 10 mL sample of a mostly neutral solution or a 100 mL sample of the same mostly neutral solution?
- What qualities should the indicators you choose for the kit have?

Tips for Evaluating the Pre-Lab Requirements

Be certain that students consider a broad range of possible substances, not just the

ones mentioned. In keeping with the true philosophy of science, the goal of the lab experience is not to have students predict accurately what substances will be useful, but rather to provide an opportunity for students to check out a variety of possibilities.

Be certain that students indicate in their plan that they will be using moderate quantities of each indicator (at least 50 mL). Plans should indicate that students will adjust the pH first to a high or low value, using acid or base and then gradually change the pH to the opposite. Although it is not necessary, many students will propose using two samples, one for testing with acid, and another for testing with base.

Proposed Procedure

Add chopped, colored food to a 400 mL beaker up to the 250 mL line. Fill the rest with water. Heat the mixture with a Bunsen burner or on a hot plate for about 15 min at a temperature of about 70°C. Allow the mixture to cool, and decant about 75 mL of the liquid into a 250 mL beaker or Erlenmeyer flask. Measure and record the color and the pH of the indicator liquid.

Set up two clean burets and fill one with 1.0 M HCl and the other with 1.0 M NaOH. Add 1.0 M HCl to the indicator, slowly. When a color change occurs, measure and record the pH and the color. Continue adding HCl until the pH is 2.0.

Using the same indicator sample, add 1.0 M NaOH gradually. Observe any color changes and record the pH and color. Continue adding NaOH until the pH is 12. Repeat the procedure with other available indicators.

Post-Lab

Disposal

Set out four disposal containers, labeled for disposal of solid wastes such as the remnants of the plant products, acidic liquid wastes, neutral liquid wastes, and basic liquid wastes. The solid wastes may be discarded in the trash can. Slowly, while stirring, combine the contents of the three liquid waste containers. Neutralize the resulting solution with 1.0 M acid or base until the pH is between 5 and 9, and then pour it down the drain.

Measuring pH—Acid Precipitation Testing

INVESTIGATION 13-2

Problem
Solving

Objectives

Students will
- use appropriate lab safety procedures.
- calculate $[H_3O^+]$ given a pH.
- perform acid-base stoichiometry calculations.
- determine how much lime will be necessary to change the pH.
- add measured quantities of lime to adjust the pH of a sample solution.
- measure pH using pH paper, a pH meter, or LEAP System with pH probe.
- model a system and extrapolate their results to a 2.5×10^9 L lake.
- design and implement their own procedure.

Planning

Recommended Time
1 lab period (can be extended)

Prerequisite
None

Materials
(for each lab group)
- CaO, lime, 0.5 g
- Distilled water
- Lake water sample (see Solution/Material Preparation)
- 150 mL beaker
- 600 mL beaker
- 100 mL graduated cylinder
- Balance
- Glass stirring rod
- pH paper, narrow range, 10 small pieces
- Spatula
- Weighing paper

pH meter option
- pH meter

LEAP System option
- LEAP System
- LEAP pH probe

Other optional equipment
- Magnetic stirrer

Estimated cost of materials:
$52 000–$62 000

Required Precautions

- Always wear goggles and a lab apron to provide protection for eyes and clothing.
- Read all safety cautions, and discuss them with your students.
- Students should not handle concentrated acid or base solutions.
- Wear goggles, a face shield, impermeable gloves, and a lab apron when you prepare the lake water solution. Work in a hood known to be in operating condition, with another person present nearby to call for help in case of an emergency. Be sure you are within 30 s walking distance of a properly working safety shower and eyewash station.
- In case of an acid or base spill, first dilute with water. Then mop up the spill with wet cloths or a wet cloth mop while wearing disposable plastic gloves. Designate separate cloths or mops for acid and base spills.
- If a magnetic stirrer, a pH meter, or the LEAP System with pH probe is used, the precautions listed in the teacher's notes on page T37 must be followed to avoid electric shock.

To prepare any volume of lake water with pH of 3.0, observe the required precautions. Add a drop of concentrated H_2SO_4 to 1.00 L of distilled water while stirring. Continue adding drops one at a time until a pH of 3.0 is reached. (It may be necessary to add small amounts of $NaHCO_3$ if the pH drops too far below 3.0.)

Student Orientation

Techniques to Demonstrate

Demonstrate the method to be used for monitoring the pH: pH meter, pH paper, or LEAP pH probe, especially any necessary standardization procedure. A magnetic stirrer is helpful in this experiment. If one is available, show how it is used.

Pre-Lab Discussion

Discuss the sources and causes of acid precipitation and the chemical reactions that oxides of nitrogen and sulfur undergo in the atmosphere. Connect the discussion to the characterization of these oxides as acidic anhydrides. Ask students to think about what regions might be most affected by acid precipitation and why regions above large limestone formations are not significantly affected.

Point out that the concentrations of acid in an acidic lake are still very small, even though they have significant biological effects. Make certain students understand that a very small amount of lime will be enough to change the pH. As a result, students should perform multiple trials. Because of its relative insolubility, students tend to add the lime too quickly and soon find that the pH is too high. If students collect data for two days, they will find the pH rises overnight as a little more of the slightly soluble CaO dissolves. Due to diffusion, this would also happen in the lake even if the water is stagnant. Assure students that although this is an inexact procedure, the objective of raising the pH enough for the survival of aquatic life can be achieved.

Tips for Evaluating the Pre-Lab Requirements

Be certain students have made an accurate prediction of the amount of CaO actually necessary, according to the calculations shown in the Sample Data and Analysis section. Students may need some prompting to

Sample Data and Analysis

pH of water = 3.0
$[H_3O^+] = 10^{(-pH)} = 1.0 \times 10^{-3}$ M

$$\text{actual molar amount} = 0.500 \text{ L} \times \frac{1.0 \times 10^{-3} \text{ mol } H_3O^+}{1 \text{ L}} = 5.0 \times 10^{-4} \text{ mol } H_3O^+$$

desired pH = 6.0
$[H_3O^+] = 10^{(-pH)} = 1.0 \times 10^{-6}$ M

$$\text{desired molar amount} = 0.500 \text{ L} \times \frac{1.0 \times 10^{-6} \text{ mol } H_3O^+}{1 \text{ L}} = 5.0 \times 10^{-7} \text{ mol } H_3O^+$$

H_3O^+ to neutralize: $5.0 \times 10^{-4} - 5.0 \times 10^{-7} = 5.0 \times 10^{-4}$ mol H_3O^+ (when rounded for significant figures)

$$5.0 \times 10^{-4} \text{ mol } H_3O^+ \times \frac{1 \text{ mol CaO}}{2 \text{ mol } H_3O^+} \times \frac{56.08 \text{ g CaO}}{1 \text{ mol}} = 1.4 \times 10^{-2} \text{ g CaO}$$

	Trial 1	Trial 2	Trial 3
Mass of 250 mL beaker + lime before use (g)	105.40	105.31	105.12
Mass of 250 mL beaker + lime after use (g)	105.37	105.29	105.10
Volume of lake water neutralized (mL)	500.0	500.0	500.0
Mass of lime used (g)	0.03	0.02	0.02

average mass of lime needed to neutralize 1 L = 0.04 g
mass of lime needed to neutralize lake = 0.04 g/L $\times$ (2.5×10^9 L) = 1.0×10^8 g = 1.0×10^5 kg

think of a way that they can keep track of the small amounts that will be added. Many will propose measuring the mass of each addition separately, but this will compound possible errors. A better approach, in which mass is measured before the procedure begins and after it is complete, is outlined in the Proposed Procedure.

Proposed Procedure

Measure and record the mass of a 150 mL beaker. Add about 1.0 g of CaO to the beaker. Measure and record the mass of the 150 mL beaker and CaO. Using a graduated cylinder, measure 500 mL of lake water into a 600 mL beaker. Measure and record the initial pH. Transfer a very small amount of CaO from the 150 mL beaker to the lake water. Stir until all of the lime is dissolved.

Monitor the pH. Continue adding lime, stirring, and measuring the pH until the pH is 6 or 7. Then measure and record the mass of the beaker with the unused CaO. Repeat for several additional trials.

Post-Lab

Disposal

Set out four disposal containers labeled for disposal of solid wastes such as any undissolved lime, acidic liquid wastes, neutral liquid wastes, and basic liquid wastes. While stirring, slowly combine the contents of the three liquid waste containers. Then, while stirring, dissolve the contents of the solid waste container in the mixture. Neutralize the resulting solution with 1.0 M acid or base until the pH is between 5 and 9, and then pour it down the drain.

Additional Notes

Acid-Base Titration— Vinegar Tampering Investigation

INVESTIGATION 13-3

Problem Solving

Objectives

Students will
- use appropriate lab safety procedures.
- measure pH with pH paper, a pH meter, or the LEAP System pH probe.
- choose an appropriate reagent for a titration.
- measure volumes using a buret or a calibrated pipet or dropper.
- titrate a sample of acetic acid in vinegar.
- perform acid-base stoichiometry calculations.
- calculate molar concentration and percentage by mass from experimental data.
- calculate $[H_3O^+]$ from pH.
- calculate the acid dissociation constant, K_a, with experimental data.
- calculate percent error using the accepted value for K_a.
- design and implement their own procedure.

Planning

Recommended Time

1 lab period

Prerequisite

Exploration 13A (from the textbook)

Materials

(for each lab group)
- 1.0 M NaOH, 60 mL
- Phenolphthalein solution, 1 mL
- Vinegar sample, 50 mL
- 150 mL beaker
- 125 mL Erlenmeyer flask, 3
- Wash bottle

Full-scale option
- Buret clamp (double)
- Buret tubes, 2
- Ring stand

Microscale option
- Pipets, 2
- 25 mL graduated cylinder

pH measurement options
- LEAP System and pH probe
- pH meter
- pH paper, narrow range, 3 pieces

Estimated cost of materials:
$63 000–$76 000

Required Precautions

- Goggles and lab apron must be worn at all times.
- Read all safety cautions, and discuss them with your students.
- Students should not handle concentrated acid or base solutions.
- Wear goggles, a face shield, impermeable gloves, and a lab apron when you prepare the vinegar samples. Wear goggles and gloves when you prepare the NaOH. Work in a hood known to be in operating condition, with another person present nearby to call for help in case of an emergency. Be sure you are within 30 s walking distance of a properly working safety shower and eyewash station.
- In case of an acid or base spill, first dilute with water. Then mop up the spill with wet cloths or a wet cloth mop while wearing disposable plastic gloves. Designate separate cloths or mops for acid and base spills.

continued on next page

Required Precautions

(continued)

- If the LEAP System with pH probe or pH meter are used, the precautions listed in the teacher's notes on page T37 must be followed to avoid electric shock.

Solution/Material Preparation

1. To prepare 1.00 L of 1.0 M NaOH, follow the required precautions. Slowly dissolve 40.0 g of NaOH in about 500 mL of distilled water while stirring. Add distilled water to bring the total volume of solution up to 1.00 L.
2. To prepare phenolphthalein solution, dissolve 1.00 g phenolphthalein in 50.0 mL of denatured alcohol, and add 50.0 mL of distilled water.
3. Purchase commercially available vinegar for samples from the plant. Be sure the label indicates 5% acidity. Pour the samples into beakers before the lab begins. If you want the students to find that the vinegar has been tampered with, follow the required precautions and add between 2–4 mL of glacial acetic acid to every 50 mL of sample.
4. Students may perform either a microscale version of this titration or a full-scale version. The most accurate results are from the full-scale procedure, but the microscale will be cheaper for the students' budgets.

Student Orientation

Techniques to Demonstrate

If you observed faulty student technique in previous titration experiments, demonstrate the process again, from rinsing the buret to filling and reading it. If microscale titration is to be used, demonstrate the use of the micropipet, particularly its calibration. Be sure to show students how to calibrate the pH meter or pH probe in a buffer solution if these tools will be used.

Pre-Lab Discussion

The pages of the textbook cited in the References section provide a fairly thorough description of the process and technique of titration. Students may have some difficulty with the calculations, especially those relating to K_a, so you may consider working through some of these calculations with sample data.

If students are uncertain about how to proceed, you may provide the following leading questions for them to consider as they plan their procedure.

- What concentrations need to be determined to calculate K_a?
- What measurements will you need to make to determine the total concentration of acetic acid in the solution?
- Will the concentration of H_3O^+ be the same as the acetic acid concentration?
- How can you measure the concentration of H_3O^+ at equilibrium?

Tips for Evaluating the Pre-Lab Requirements

Students should plan to do at least three trials for the titration and pH measurement steps. A 15 mL sample size for each trial is appropriate. Be certain students' plans indicate that they understand that the total acetic acid concentration derived from the titration portion of the lab cannot be used directly in the calculations for the K_a for the original solution. The concentration of ionized acid will be the same as the H_3O^+ concentration, and the concentration of un-ionized acid will be the total concentration minus the ionized concentration.

Make certain student plans' recognize that the end point of the titration is the very first persistent appearance of a pale pink color, not the vivid magenta that appears after too much NaOH has been added.

Proposed Procedure

Measure and record the pH of the original vinegar sample. If a buret is being used, clean and rinse it with 1.0 M NaOH. Then fill it with 1.0 M NaOH, making sure that the tip is full, and record the initial reading. If a microscale titration is being done, calibrate two pipets, one for vinegar and one for NaOH, and then rinse each with the appropriate solution. If buret is being used, measure a 10–15 mL sample of vinegar into a 125 mL Erlenmeyer flask, and add a few drops of phenolphthalein. If a microscale titration is being done, add a specific number of drops of vinegar with a pipet. Titrate the vinegar with NaOH until the appearance of the first persistent pink color. Record the final volume of NaOH used. Repeat the titration two times.

Post-Lab

Disposal

Set out three disposal containers labeled for disposal of acidic liquid wastes, neutral liquid wastes, and basic liquid wastes. While stirring, slowly combine the contents of the three liquid-waste containers. Neutralize the resulting solution with 1.0 M acid or base until the pH is between 5 and 9, and then pour it down the drain.

Sample Data and Analysis

pH of vinegar sample = 2.50
$[H_3O^+] = \log^{(-pH)} = 3.16 \times 10^{-3}$

Buret Titration

Trial	Vinegar initial (mL)	Vinegar final (mL)	NaOH initial (mL)	NaOH final (mL)
1	0.55	13.20	1.38	12.56
2	13.20	27.43	12.56	24.65
3	27.43	40.67	24.65	36.52

Trial 1:

$$0.01118 \text{ L NaOH} \times \frac{1.0 \text{ mol NaOH}}{1 \text{ L}} \times \frac{1 \text{ mol CH}_3\text{COOH}}{1 \text{ mol NaOH}} \times \frac{1}{12.65 \text{ mL}} \times \frac{1000 \text{ mL}}{1 \text{ L}} =$$

$$0.8838 \text{ M CH3COOH}$$

Trial 2:

$$0.01209 \text{ L NaOH} \times \frac{1.0 \text{ mol NaOH}}{1 \text{ L}} \times \frac{1 \text{ mol CH}_3\text{COOH}}{1 \text{ mol NaOH}} \times \frac{1}{14.20 \text{ mL}} \times \frac{1000 \text{ mL}}{1 \text{ L}} =$$

$$0.8514 \text{ M CH}_3\text{COOH}$$

Trial 3:

$$0.01187 \text{ L NaOH} \times \frac{1.0 \text{ mol NaOH}}{1 \text{ L}} \times \frac{1 \text{ mol CH}_3\text{COOH}}{1 \text{ mol NaOH}} \times \frac{1}{13.24 \text{ mL}} \times \frac{1000 \text{ mL}}{1 \text{ L}} =$$

$$0.8965 \text{ M CH}_3\text{COOH}$$

Average = 0.8772 M CH$_3$COOH

$$K_a = \frac{[CH_3COO^-][H_3O^+]}{[CH_3COOH]} = \frac{(3.16 \times 10^{-3})^2}{0.8772 - (3.16 \times 10^{-3})} = 1.14 \times 10^{-5}$$

$$\frac{0.8772 \text{ mol CH}_3\text{COOH}}{1 \text{ L}} \times \frac{60.06 \text{ g CH}_3\text{COOH}}{1 \text{ mol}} \times \frac{1 \text{ L}}{1010 \text{ g solution}} \times 100 = 5.216\% \text{ by mass}$$

Microscale Titration

Trial	Drops of vinegar	Drops of NaOH
1	100	110
2	75	84
3	50	57

$$\frac{2.5 \text{ mL}}{60 \text{ drops}} = 0.042 \text{ mL/drop}$$

(Once volumes are converted to milliliters, the calculations should follow the same pattern.)

Catalysts

Objectives

Students will
- use appropriate lab safety procedures.
- observe the decomposition of hydrogen peroxide.
- measure mass using a laboratory balance.
- measure volume using a graduated cylinder and a buret.
- measure the time needed for the production of a given volume of product using a stopwatch or clock with a second hand.
- graph experimental data and calculate the slope of a graph.
- relate the reaction rate to the slope of a graph of volume of product vs. time.
- infer the effect of a catalyst on the rate of the reaction.
- determine the effect of the concentration of hydrogen peroxide on the reaction rate.

Planning

Recommended Time

1 lab period

Materials

(for each lab group)
- 3% H_2O_2 solution, 125 mL
- Distilled water
- NaI, 0.50 g
- 125 mL Erlenmeyer flask
- 100 mL graduated cylinder
- 100 mL Mohr buret
- Balance
- Buret clamp
- Leveling bulb
- Medicine droppers, 2
- Pneumatic trough or plastic tub
- Ring stand and ring
- Rubber tubing, 2 pieces, about 20 cm each
- Rubber stopper, one-hole (for flask)
- Rubber stopper, one-hole (for Mohr buret)
- Stopwatch or clock with second hand
- Thermometer, nonmercury
- Weighing paper, 3 pieces

LEAP System option
- LEAP System
- LEAP thermistor probe

Solution/Material Preparation

1. Use only 3% H_2O_2 for this Exploration. This is the concentration most commonly found in pharmacies and other stores. Do not use more concentrated forms, and do

Required Precautions

- Goggles and a lab apron must be worn at all times to provide protection for eyes and clothing.
- Read all safety cautions, and discuss them with your students.
- If the LEAP System with thermistor probe is used, the precautions listed in the teacher's notes on page T37 must be followed to avoid electric shock.

not try to dilute more concentrated forms to make 3% H_2O_2. To make sure your supply is reasonably fresh, test it by pouring a little into a small beaker. Add a pinch of NaI. If there is no bubbling present, the hydrogen peroxide has already decomposed, and you will need a new supply.

2. Before the lab, prepare several sets of medicine droppers for use. First, put on cloth gloves and wrap the medicine droppers in a layer of cloth. Insert them into the one-hole rubber stoppers. It may help to apply a small amount of lubricant to the sides of the droppers.

3. Note that the top of the leveling bulb must be open in order for the apparatus to function properly.

Student Orientation

Techniques to Demonstrate

Show students how to assemble the apparatus and use the leveling bulb. It is very important that they follow the directions to check the apparatus for leaks.

Pre-Lab Discussion

Students may need help in understanding the function of the leveling bulb. Discuss the need for the volumes of oxygen to be measured at constant pressure. Proper use of the leveling bulb makes it possible for all volumes to be measured at the same pressure. When the water levels are the same, the pressure on the water outside of the bulb, which is the atmospheric pressure, is the same as the pressure of the trapped gas. If the water level was higher in the bulb, the pressure of the gas would be different.

Remind students that rate is simply a measure of change in a product or reactant amount in a given amount of time. Perform some sample rate calculations before the lab so that students will understand what they are measuring and why.

Discuss with students why they add the hydrogen peroxide after the other chemicals have been mixed and adjusted to the temperature of the water bath.

Technically, this lab is very difficult to perform. Emphasize that it is important that both lab partners know what they are responsible for before they begin the reaction.

Post-Lab

Disposal

Set out a disposal container for any leftover solid NaI. Students are instructed to pour any leftover solutions down the drain once they have stopped bubbling. The NaI may be either dissolved and poured down the drain or saved and used for the next time this Exploration is performed.

Answers to

Analysis and Interpretation

1. The five factors affecting rate are the nature of reactants, particle size, temperature, concentration, and presence of catalysts. In this experiment, concentration and the influence of a catalyst were studied.

2.

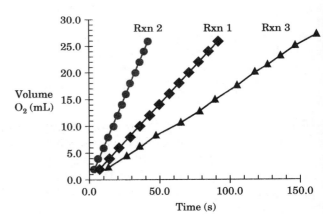

3. The slopes of the three lines can be calculated using the first and last points.

   Rxn 1: slope $= \dfrac{24}{86} = 0.28$ mL O_2/s

   Rxn 2: slope $= \dfrac{24}{40} = 0.60$ mL O_2/s

   Rxn 3: slope $= \dfrac{24}{160} = 0.15$ mL O_2/s

Sample Data and Analysis

	Rxn 1 (s)	Rxn 2 (s)	Rxn 3 (s)
NaI (g)	0.10	0.20	0.10
3% H_2O_2 (mL)	50.0	50.0	25.0
H_2O (mL)	0.0	0.0	25.0
2 mL O_2	7	3	13
4 mL O_2	14	6	25
6 mL O_2	21	10	37
8 mL O_2	29	13	50
10 mL O_2	36	17	62
12 mL O_2	43	20	74
14 mL O_2	50	23	86
16 mL O_2	57	26	98
18 mL O_2	64	29	111
20 mL O_2	71	33	123
22 mL O_2	78	36	135
24 mL O_2	85	39	147
26 mL O_2	93	43	160

The slope is equal to the rate for each of the three reactions.

4. Keeping the bulb's water level even with the water level of the buret ensured that the oxygen gas trapped in the buret was at the same pressure as the atmosphere. The temperature and pressure had to be kept constant so that the different volumes of O_2 could be compared.

Conclusions

5. Within experimental error, the rate doubled from Reaction 2 to Reaction 1, when the concentration of iodide ion was doubled.

6. The concentration of H_2O_2 in Reaction 1 was double the concentration in Reaction 3. Although the rate of Reaction 1 is not exactly double the rate of Reaction 3, it is very close. In other words, there is a linear relationship between rate and $[H_2O_2]^1$.

7. rate $= k[I^-]^1[H_2O_2]^1$

8. At constant temperature and pressure, a given volume of gas contains a definite number of particles. Therefore, increases in volume relate directly to increases in number of particles as more gas is added. This is equivalent to increases in molar concentration in aqueous systems.

9. When the concentration of a reactant is increased, there will often be more frequent collisions between reacting particles, and these will result in an increase in the rate.

10. When a reactant is not a part of the rate-determining step of a reaction, increasing the concentration of that reactant has no effect on the rate.

11. Student answers will vary. Given that the volume is about 20 times as much, students should suggest using at least 20 times as much of each reactant as was used in Reaction 2. Some may suggest using even more NaI.

Extensions

1. Students' suggestions for improving the procedure will vary. For most students, experimental error caused by unfamiliarity with the equipment will be the largest error. Repeated trials should provide better data. Be sure answers are safe and include carefully planned procedures.

2. For each period between measurements, the average rate $= \Delta mL\ O_2/\Delta t$.

first period: $\dfrac{2.0\ \text{mL}\ O_2}{60.0\ \text{s}} = 0.033\ \text{mL}\ O_2/\text{s}$

second period: $\dfrac{1.8\ \text{mL}\ O_2}{55.0\ \text{s}} =$

$0.033\ \text{mL}\ O_2/\text{s}$

third period: $\dfrac{2.1\ \text{mL}\ O_2}{65.0\ \text{s}} =$

$0.032\ \text{mL}\ O_2/\text{s}$

fourth period: $\dfrac{2.0\ \text{mL}\ O_2}{65.0\ \text{s}} =$

$0.032\ \text{mL}\ O_2/\text{s}$

3. After another period of about 2 s, the rate might be 0.031 mL O_2/s because the rate is decreasing with the progress of the reaction.

4. A catalyst provides a lower-energy pathway or mechanism for a reaction. Lower activation energy means that at a given temperature more particles will possess the necessary energy to react.

5. In areas where dry, combustible, powdered materials are in the air, there is danger of explosion because these materials have a large surface area in contact with atmospheric oxygen. The presence of a spark can initiate a very rapid reaction.

6. Students will find that the body uses enzyme catalysts such as catalase and peroxidase to decompose toxic hydrogen peroxide. Textbooks on biochemistry are good sources of information.

7. Student answers will vary. Students should probably select the peroxide reaction over the hydrochloric acid one because hydrochloric acid is caustic and could cause a burn if spilled. Also, the product, hydrogen gas, is flammable.

Catalysts—Peroxide Disposal

INVESTIGATION 14-1

Problem Solving

Objectives

Students will
- use appropriate lab safety procedures.
- observe the decomposition of hydrogen peroxide.
- measure mass using a laboratory balance.
- measure volume using a graduated cylinder and a buret.
- measure the time needed for the production of a given volume of product using a stopwatch or clock with a second hand.
- graph experimental data.
- calculate the slope of a graph.
- relate the reaction rate to the slope of a graph of volume of product vs. time.
- compare the rates for the reaction using three different catalysts.
- evaluate which catalyst is most effective and cost-efficient.
- design and implement their own procedure.

Planning

Recommended Time
2 lab periods (test some catalysts on second day)

Prerequisite
Exploration 14-1

Materials
(for each lab group)
- 3% H_2O_2, 50 mL
- NaI, 1 g
- MnO_2, 1 g
- Yeast, 1 g
- 100 mL Mohr buret
- 125 mL Erlenmeyer flask
- 100 mL graduated cylinder
- Balance
- Buret clamp
- Leveling bulb
- Medicine droppers, 2
- Pneumatic trough or plastic tub
- Ring stand and ring
- Rubber stopper, one-hole (for Erlenmeyer flask)
- Rubber stopper, one-hole (for Mohr buret)
- Rubber tubing, 2 pieces, about 20 cm each
- Stopwatch or clock with second hand
- Thermometer, nonmercury
- Weighing paper, 3 pieces

LEAP System option
- LEAP System
- LEAP thermistor probe

Estimated cost of materials:
$175 000–$181 000

Required Precautions

- Goggles and a lab apron must be worn at all times to provide protection for eyes and clothing.
- Read all safety cautions, and discuss them with your students.
- No burners, flames, hot plates, or other heat sources should be in use in the lab because the oxygen being generated can support combustion better than air.
- Be certain the lab room is well ventilated.
- If the LEAP System with thermistor probe is used, the precautions listed in the teacher's notes on page T37 must be followed to avoid electric shock.

Sample Data and Analysis

Vol. O$_2$ (mL)	Rxn 1, NaI (s)	Rxn 2, MnO$_2$ (s)	Rxn 3, yeast (s)
2	7	10	1
4	14	20	2
6	21	31	3
8	29	42	4
10	36	52	5
12	43	62	6
14	50	71	7
16	57	82	8
18	64	91	9
20	71	101	10
22	78	111	11
24	85	122	12
26	92	132	13

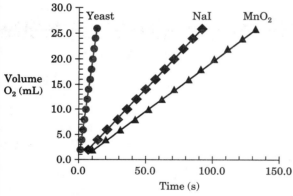

Students' answers about the rate of each reaction will vary depending on the many variables in this procedure. Because of the complexities involved, there is no clear-cut, best answer to this problem. NaI is faster than MnO$_2$, but it is so much more expensive that NaI probably isn't the most cost-effective one to use. Both of these inorganic catalysts can be recovered by evaporation and reused. Yeast is the cheapest catalyst and the fastest. The disadvantage of yeast is that it may be difficult to keep alive. Accept any student answer that makes a convincing case for the catalyst chosen. (Note that the costs cited are in proportion to the actual cost of these catalysts as reagents.)

Solution/Material Preparation

1. Use only 3% H$_2$O$_2$ for this Exploration. This is the concentration most commonly found in pharmacies and other stores. Do not use more concentrated forms, and do not try to dilute more concentrated forms to make 3% H$_2$O$_2$. To make sure your supply is reasonably fresh, test it by pouring a little into a small beaker. Add a pinch of NaI. If there is no bubbling present, the hydrogen peroxide has already decomposed, and you will need a new supply.

2. Before the lab, prepare several sets of medicine droppers for use. First, put on cloth gloves, and wrap the medicine droppers in a layer of cloth. Insert them into the one-hole rubber stoppers. It may help to apply a small amount of lubricant to the side of the droppers.

3. Note that the top of the leveling bulb must be open in order for the apparatus to function properly.

Student Orientation

Techniques to Demonstrate

Show students how to assemble the apparatus and how to use the leveling bulb. It is very important that they follow the directions given in the lab for checking the apparatus for leaks.

Pre-Lab Discussion

Begin by discussing the results and procedure used in the Exploration. If students are not certain about how to proceed, encourage them to use quantities similar to those in the Exploration.

Be sure to discuss the difference between the three catalysts. Sodium iodide, NaI, is a soluble alkaline halide. Manganese dioxide, MnO$_2$, is an insoluble oxide of a transition metal. Yeast is a living organism, a fungus that catalyzes the decomposition of H$_2$O$_2$ enzymatically. Students may be concerned about the insolubility of manganese dioxide. Tell them that this fact may influence the advice they give to their client.

Students will be uncertain about how much of each catalyst to order. Remind students that these catalysts are not consumed by a reaction, so a little can go a long way. If there is time, encourage students to perform multiple trials.

Be certain students have ordered only a small amount of catalyst. The procedure can be performed with as little as 0.20 g, but students may order as much as 1.0 g. If they suggest using more than 1.0 g, have them reconsider their plans. To properly evaluate each possibility, they should buy the same mass of each catalyst.

Proposed Procedure

Assemble the buret, pneumatic trough, flask, and leveling bulb. Check the system for leaks, and adjust the temperature of the water to room temperature. Measure about 0.20 g of each of three catalysts, and record the mass to the nearest 0.01 g. Record the initial volume of the buret. Place the first catalyst in the flask. Measure about 50.0 mL of 3% H_2O_2 into a graduated cylinder, and record the volume to the nearest 0.1 mL. Add the H_2O_2 to the flask, and quickly attach the flask to the buret with the stopper-medicine-dropper-tubing assembly. Record the time and volume of oxygen produced every 2 mL. Take about 13 readings. Check the temperature of the water, and adjust it to room temperature, if necessary. Repeat the procedure two times, using MnO_2 and yeast as catalysts.

Post-Lab

Disposal

Set out three disposal containers. The H_2O_2 solution with MnO_2 powder in it and any leftover MnO_2 solids should be placed in the first one. Leftover solid NaI and yeast should be placed in the other two containers, respectively. Any H_2O_2, NaI, or yeast solutions may be poured down the drain. **Do not wash MnO_2 down the drain.** The MnO_2 solutions should be evaporated, and, if the solutions were not contaminated, the recovered MnO_2 can be reused. Otherwise, the recovered MnO_2 can be placed in the trash can to be sent to a landfill.

Additional Notes

Electroplating for Corrosion Protection

EXPLORATION

15-1
Technique Builder

Objectives

Students will
- use appropriate lab safety procedures.
- construct an electrolytic cell and use it to plate metals.
- compare the masses of iron and zinc electroplated onto a copper wire.
- test how plated metals resist corrosion by an acidic solution.
- write balanced chemical equations for reactions at each electrode.
- relate the results to the activity series and to reduction potentials.

Planning

Recommended Time

1 lab period with drying oven; 2 lab periods otherwise

Materials

(for each lab group)
- 1.0 M HCl, 50 mL
- $FeCl_3$ plating solution, 80 mL
- $ZnSO_4$ plating solution, 80 mL
- Copper wire, 10 cm lengths, 3
- Distilled water
- Iron strip, 1 cm × 8 cm
- Zinc strip, 1 cm × 8 cm
- 150 mL beaker, 3
- 400 mL beaker
- Balance
- Battery, 6 V (lantern-type)
- Steel wool
- Stick-on label
- Stopwatch or clock with second hand
- Test-tube rack
- Test tubes, large, 3
- Wax pencil
- Wire with alligator clips, 2 pieces

Optional equipment
- Beaker tongs
- Drying oven

Solution/Material Preparation

I. To prepare 1.00 L of the $FeCl_3$ plating solution, observe the required precautions. Dissolve 40.0 g $FeCl_3 \cdot 6H_2O$ in 900 mL of distilled water and 50 mL of ethanol. Add more water to dilute to 1.00 L.

Required Precautions

- Goggles and a lab apron must be worn at all times to provide protection for eyes and clothing.
- Read all safety cautions, and discuss them with your students.
- Remind students that objects from the drying oven will be very hot. They should use beaker tongs or a hot mitt to pick them up.
- Students should not handle concentrated acid or base solutions.
- Wear goggles, a face shield, impermeable gloves, and a lab apron when preparing the HCl. Work in a hood known to be in operating condition, with another person present nearby to call for help in case of an emergency. Be sure you are within 30 s walking distance of a properly working safety shower and eyewash station.
- In case of an acid or base spill, first dilute with water. Then mop up the spill with wet cloths or a wet cloth mop while wearing disposable plastic gloves. Designate separate cloths or mops for acid and base spills.
- Observe the precautions on the ethanol bottle's label when preparing the $FeCl_3$ solution.
- If a drying oven is used, the precautions listed on page T37 must be followed to avoid electric shock.

2. To prepare 1.00 L of the $ZnSO_4$ plating solution, dissolve 50.0 g $ZnSO_4 \cdot 7H_2O$, 24.0 g NH_4Cl, and 40.0 g ammonium citrate, $(NH_4)_2HC_6H_5O_7$, in 900 mL of distilled water. Add more water to dilute to 1.00 L.

3. To prepare 1.00 L of 1.00 M HCl, observe the required precautions. Add 82.6 mL of concentrated HCl to enough distilled water to make 1.00 L of solution. Add the acid slowly, and stop frequently to stir it in order to avoid overheating.

4. Any size of copper wire will do. The procedure that generated the sample data was performed with soft, bare, 18 gauge wire.

5. Use a 6 V "lantern battery" or its equivalent. DO NOT use a high-amperage 6 V battery such as those designed for supplying current for the spark in gasoline-powered motors.

6. Although optional, the use of the drying oven is strongly recommended. Otherwise, if the lab is to be performed quantitatively, the wires must dry overnight after plating and again after testing them in 1.0 M HCl.

Student Orientation

Techniques to Demonstrate

Be sure to show students the proper order for the connections between the electrodes and the battery, reminding them to check that they have connected the appropriate

Sample Data and Analysis

	Cu	Fe/Cu	Zn/Cu
Mass of wire (g)	0.77	0.77	0.77
Plating time (s)	n/a	301	299
New mass of wire (g)	0.77	0.82	0.89
HCl time (s)	300	300	300
Mass of wire after HCl (g)	0.77	0.77	0.77

Students' data should also include observations about the wire. After plating, the metals should be visible on the wire. The iron will look dark and silvery, and the zinc will look silvery, but lighter than the iron. After treatment with HCl, little of the plated metals will remain.

metal to the battery's positive electrode. Point out the importance of performing the steps in exactly the same manner for the different wire treatments. Each wire should be plated for the same amount of time, at the same depth in the beaker, and with similar amounts of plating solution. Similarly, the testing steps should be performed with the same amounts of acid for the same amount of time.

Pre-Lab Discussion

Discuss reduction and oxidation. Make sure students realize that a different reaction is occurring at each electrode. Many will recognize that plating (reduction) is occurring at one electrode but will not understand that oxidation of the metal is occurring at the other.

Post-Lab

Disposal

Set out six disposal containers: three bins and three bottles, or other similar containers. Of the three bins, use one for copper wires, one for zinc metal strips, and one for iron metal strips. Of the three containers, use one for the $FeCl_3$ solution, one for the $ZnSO_4$ solution, and one for the contents of the Waste beaker and the HCl from the test tubes.

The metal strips should be cleaned with soap and water, rinsed, and dried for reuse the next time students perform this Exploration. The copper wires can also be reused after they have been treated with 1.0 M HCl and all of the zinc or iron has been removed. The iron and zinc solutions should be saved and reused.

Pour the HCl that has been used in treating the copper wires into the disposal container for the Waste beaker. Add 1.0 M NaOH to the mixture in the Waste beaker container to precipitate the iron and zinc as hydroxides. Filter the mixture, placing the precipitate in the trash. Then neutralize the filtrate with 1.0 M H_3PO_4 until the pH is between 5 and 9, and pour it down the drain.

Answers to

Analysis and Interpretation

1. iron strip: $Fe(s) \longrightarrow Fe^{3+}(aq) + 3e^-$, anode
Copper wire: $Fe^{3+}(aq) + 3e^- \longrightarrow Fe(s)$, cathode

zinc strip: $Zn(s) \longrightarrow Zn^{2+}(aq) + 2e^-$,
anode
copper wire: $Zn^{2+}(aq) + 2e^- \longrightarrow Zn(s)$,
cathode

2. 0.05 g Fe, 0.12 g Zn
   9×10^{-4} mol Fe, 1.8×10^{-3} mol Zn

Conclusions

3. The untreated copper wire was least reactive. When the zinc- and iron-plated wires were placed in the HCl, bubbling occurred, indicating a reaction.

4. Students' suggestions about the disadvantages of copper will vary. Possible answers include the following: copper is soft and malleable, so the tanks might be easily harmed; copper costs more than iron or zinc.

Extensions

1. Students will find a variety of methods in use. The Teflon used in the Statue of Liberty, as described in Chapter 7, prevents the iron and copper in the statue from coming into direct contact. In this way, the redox reactions that led to the crumbling and corrosion of the previous iron framework can be greatly reduced. Other coatings and paints work the same way to keep the reactants separated. In other approaches, a sacrificial anode, which is made of a more active metal than the main part of the item, is used. This approach is used on some ships' hulls. Another possibility is to apply a small amount of voltage to an object so that reduction rather than oxidation is favored.

Additional Notes

Voltaic Cells

EXPLORATION

15-2
Technique Builder

Objectives

Students will
- use appropriate lab safety procedures.
- design and construct various voltaic cells.
- measure the actual voltages of the voltaic cells using a voltmeter or the LEAP System with voltmeter probe.
- calculate the voltages of the voltaic cells using a table of standard reduction potentials.
- evaluate cells by comparing actual cell voltages to voltages calculated from standard reduction potentials.
- write equations for cell reactions by combining half-cell reactions.

Planning

Recommend Time

1 or 2 lab periods (test different cells on different days)

Equipment/Materials

(for each lab group)
- 0.5 M $Al_2(SO_4)_3$, 75 mL
- 0.5 M $CuSO_4$, 75 mL
- 0.5 M $ZnSO_4$, 75 mL
- Aluminum strip, 1 cm $\times$ 8 cm
- Copper strip, 1 cm $\times$ 8 cm
- Zinc strip, 1 cm $\times$ 8 cm
- Distilled water
- 100 mL graduated cylinder
- Balance
- Emery cloth

Porous cup option
- 150 mL beakers, 3
- 250 mL beakers, 3
- Porous cup

Salt bridge option
- 150 mL beakers, 3
- Salt bridge

Voltmeter option
- Voltmeter
- Wires with alligator clips, 2

LEAP System option
- LEAP System
- LEAP voltage probe

Solution/Material Preparation

Note that exact molarities are not necessary for the success of this lab.

1. To prepare 1.00 L of 0.5 M $ZnSO_4$, dissolve 144 g of $ZnSO_4 \cdot 7H_2O$ in about

Required Precautions

- Goggles and a lab apron must be worn at all times to provide protection for eyes and clothing.
- Read all safety cautions, and discuss them with your students.
- If the LEAP System with voltage probe or an electric voltmeter is used, the precautions listed in the teacher's notes on page T37 must be followed to avoid electric shock.

500 mL of distilled water while stirring. Add more distilled water to bring the total volume of solution up to 1.00 L.

2. To prepare 1.00 L of 0.5 M $CuSO_4$, dissolve 125 g of $CuSO_4 \cdot 5H_2O$ in 500 mL of distilled water while stirring. Add more distilled water to bring the total volume of solution up to 1.00 L.

3. To prepare 1.00 L of 0.5 M $Al_2(SO_4)_3$, dissolve 297 g of $Al_2(SO_4)_3 \cdot 14H_2O$ in 500 mL of distilled water while stirring. Add more distilled water to bring the total volume of solution up to 1.00 L.

4. If students prepare their own 75 mL solutions, they will need about 11 g of $ZnSO_4 \cdot 7H_2O$, 9 g of $CuSO_4 \cdot 5H_2O$, and 22 g of $Al_2(SO_4)_3 \cdot 14H_2O$. These amounts are for solutions of approximate molarities; exact molarities are not necessary.

5. Porous cups should be soaked in water for several hours prior to use.

6. To make your own salt bridges, bend glass tubing into ∪ shapes. Fill the tubes with 1.0 M NaCl solution, and plug the ends with cotton. Voltages will be lower with salt bridges than with porous cups. An even simpler approach is to roll up a piece of thick blotter paper or large filter paper that has been soaked until dripping with the salt solution. Drape one end into one beaker, and the other end into the other beaker.

7. To prepare 100 mL of 1.0 M NaCl for the salt bridges, dissolve 5.8 g NaCl in enough distilled water make 100 mL of solution.

8. The LEAP voltage probe is merely a pair of wires with a phone-cord link at one end and separate alligator clips at the other. The LEAP System box, into which probes are plugged, measures the potential differences as a part of its analog-to-digital signal conversion.

Student Orientation

Techniques to Demonstrate

Have a voltaic cell like those the students will construct set up and ready to attach to a voltmeter or LEAP System. Demonstrate what happens when the cell is connected with the electron flow in the wrong direction, and show how to reverse the leads.

Pre-Lab Discussion

Review oxidation-reduction reactions and their relationship to voltaic cells. Use an example to show how voltaic cell design provides a means of separating the two redox processes into separate half-cells. Use drawings and diagrams because some students have difficulty with oral descriptions of electrochemical reactions.

Review writing equations for half-cell reactions and determining cell voltages from standard reduction potentials. Be certain

students realize that rewriting a reduction half-reaction as an oxidation requires a change in sign. Define the terms *reducing agent* and *oxidizing agent*.

Have students predict the voltages of the cells they will construct prior to beginning the lab work.

Post-Lab

Disposal

Set out six labeled disposal containers, one for each metal and solution. The metals should be saved and reused. The solutions can be evaporated, and the dry sulfates can be saved for reuse. Even with the possibility of small amounts of contamination, they are acceptable for use the next time students perform this Exploration. **Do not pour these solutions down the drain. Do not place these metals, the solutions, or the dry sulfates in the trash can.**

Answers to

Analysis and Interpretation

1. $Zn(s) \longrightarrow Zn^{2+}(aq) + 2e^-$
    $E^0 = +0.76$ V
    $Cu^{2+}(aq) + 2e^- \longrightarrow Cu(s)$
    $E^0 = +0.34$ V

    $Zn^{2+}(aq) + 2e^- \longrightarrow Zn(s)$
    $E^0 = -0.76$ V
    $Al(s) \longrightarrow Al^{3+}(aq) + 3e^-$
    $E^0 = +1.66$ V

    $Al(s) \longrightarrow Al^{3+}(aq) + 3e^-$
    $E^0 = +1.66$ V
    $Cu^{2+}(aq) + 2e^- \longrightarrow Cu(s)$
    $E^0 = +0.34$ V

2. $Zn(s) + Cu^{2+}(aq) \longrightarrow Zn^{2+}(aq) + Cu(s)$
    $2Al(s) + 3Zn^{2+}(aq) \longrightarrow$
    $\qquad\qquad 2Al^{3+}(aq) + 3Zn(s)$
    $2Al(s) + 3Cu^{2+}(aq) \longrightarrow$
    $\qquad\qquad 2Al^{3+}(aq) + 3Cu(s)$

Sample Data and Analysis			
Cell	Zn/Zn²⁺//Cu²⁺/Cu	Al/Al³⁺//Zn²⁺/Zn	Al/Al³⁺//Cu²⁺/Cu
Diagram (See item 8. Student diagrams will vary, but should clearly indicate the metals.)			
Conc.	0.5 M Zn²⁺	0.5 M Al³⁺	0.5 M Al³⁺
	0.5 M Cu²⁺	0.5 M Zn²⁺	0.5 M Cu²⁺
Voltage (V)	0.96	0.83	1.90

3. E^0 (Zn/Cu) = + 0.76 V + 0.34 V = + 1.10 V

E^0 (Al/Zn) = − 0.76 V + 1.66 V = + 0.90 V

E^0 (Al/Cu) = + 1.66 V + 0.34 V = + 2.00 V

4. Students will measure lower voltages than the standard electrode potentials because they were not using 1.0 M solutions. The porous cup or salt bridge is not a perfect conductor, and polarization of the electrodes may occur.

Conclusions

5. If the cells are constructed with care, the Al/Cu cell should be the one recommended to NASA.

6. Aluminum is the strongest reducing agent of the three metals. It is oxidized by Zn and Cu. Copper is the strongest oxidizing agent. It is reduced by Al and Zn.

7. An oxide coating on a metal electrode could prevent the ions in solution from coming into contact with the metal.

8.

Extensions

1. If the cells were left connected, they would continue reacting until all of the metals that were being oxidized were consumed.

2. Aluminum is being oxidized in two of the cells. Because its molar mass is so much less than zinc's molar mass, the Al/Cu cell is probably the longest-lasting, even after stoichiometric considerations are taken into account.

3. Students' answers will vary. From the results of the lab, it can be seen that the Al/Cu cell has the same potential as the Al/Zn cell plus the Zn/Cu cell. Students should realize that if they connect cells in series like this, they can increase the voltage. Be sure answers are safe and include carefully planned procedures.

4. In the galvanic cell pictured, manganese will be oxidized.

5. Students' answers will vary. Students should note that although electric cars do not emit pollutants as they operate, the process of generating electricity for the cars often causes damage to the environment. Similarly, many of the best options for long-term storage batteries involve hazardous chemicals.

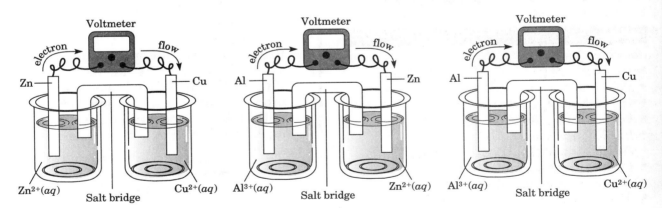

Voltaic Cells—
Designing Batteries

INVESTIGATION 15-2

Problem Solving

Objectives

Students will
- use appropriate lab safety procedures.
- design and construct various voltaic cells.
- measure the actual voltages of the voltaic cells using a voltmeter or the LEAP System with voltmeter probe.
- calculate the voltages of the voltaic cells using a table of standard reduction potentials.
- evaluate cells by comparing actual cell voltages to voltages calculated from standard reduction potentials.
- determine the effect of connecting cells in series.
- create a set of cells that will provide a given voltage.
- write equations for cell reactions by combining half-cell reactions.
- design and implement their own procedure.

Planning

Recommended Time
1 or 2 lab periods for thorough investigation

Prerequisite
Exploration 15-2

Materials
(for each lab group)*
- 0.5 M $Al_2(SO_4)_3$, 150 mL
- 0.5 M $CuSO_4$, 150 mL
- 0.5 M $MgSO_4$, 150 mL
- 0.5 M $SnCl_2$, 150 mL*
- 0.5 M $ZnSO_4$, 150 mL*
- Aluminum strips, 1 cm × 8 cm, 2
- Copper strips, 1 cm × 8 cm, 2
- Magnesium strips, 1 cm × 8 cm, 2
- Tin strips, 1 cm × 8 cm, 2*
- Zinc strips, 1 cm × 8 cm, 2*
- Distilled water
- Emery cloth

Porous cup option
- 150 mL beakers, 5
- 250 mL beakers, 5
- Porous cups, 3

Salt bridge option
- 150 mL beakers, 5
- Salt bridges, 3

Voltmeter option
- Voltmeter
- Wires with alligator clips, 4

Required Precautions

- Goggles and a lab apron must be worn at all times to provide protection for eyes and clothing. Gloves must be worn by anyone who will work with the $SnCl_2$ solution.
- Read all safety cautions, and discuss them with your students.
- If the LEAP System with voltage probe is used, the precautions listed in the teacher's notes on page T37 must be followed to avoid electric shock.

LEAP system option
- LEAP System
- LEAP voltage probe
- Wires with alligator clips, 2

* Some students may recognize that the items marked here with an asterisk are unlikely to be useful for high-voltage cells, but it is still recommended that you have all metals and solutions available so that students are forced to think through their choices.

Estimated cost of materials:
  $95 000–$110 000

Solution/Material Preparation

Note that exact molarities are not necessary for the success of the lab.

1. To prepare 1.00 L of 0.5 M $MgSO_4$, dissolve 123 g of $MgSO_4 \cdot 7H_2O$ in about 500 mL of distilled water while stirring. Add distilled water to bring the total volume of solution up to 1.00 L.

2. To prepare 1.00 L of 0.5 M $CuSO_4$, dissolve 125 g of $CuSO_4 \cdot 5H_2O$ in 500 mL of distilled water while stirring. Add distilled water to bring the total volume of solution up to 1.00 L.

3. To prepare 1.00 L of 0.5 M $Al_2(SO_4)_3$, dissolve 297 g of $Al_2(SO_4)_3 \cdot 14H_2O$ in 500 mL of distilled water while stirring. Add distilled water to bring the total volume of solution up to 1.00 L.

4. To prepare 1.00 L of 0.5 M $ZnSO_4$, dissolve 144 g of $ZnSO_4 \cdot 7H_2O$ in about 500 mL of distilled water while stirring. Add distilled water to bring the total volume of solution up to 1.00 L.

5. To prepare 1.00 L of 0.5 M $SnCl_2$, wear gloves and dissolve 113 g of $SnCl_2 \cdot 2H_2O$ in about 500 mL of distilled water while stirring. If the solution turns cloudy, slowly add a few milliliters of 1.0 M HCl while stirring. Add distilled water to bring the total volume of solution up to 1.00 L. Note that the $SnCl_2$ solution irritates the skin.

6. If students prepare their own 150 mL solutions, they will need about 19 g of $MgSO_4 \cdot 7H_2O$, 19 g of $CuSO_4 \cdot 5H_2O$, 44 g of $Al_2(SO_4)_3 \cdot 14H_2O$, 22 g of $ZnSO_4 \cdot 7H_2O$, and 17 g of $SnCl_2 \cdot 5H_2O$. These amounts are for solutions of approximate molarities; exact molarities are not necessary.

7. Porous cups should be soaked in water for several hours prior to use.

8. To make your own salt bridges, bend glass tubing into ∪ shapes. Fill the tubes with 1.0 M NaCl solution, and plug the ends with cotton. Voltages will be lower with salt bridges than with porous cups. An even simpler approach is to roll up a piece of thick blotter paper or large filter paper that has been soaked until dripping with the salt solution. Drape one end into one beaker, and the other end into the other beaker.

9. To prepare 100 mL of 1.0 M NaCl for the salt bridges, dissolve 5.8 g NaCl in enough distilled water make 100 mL of solution.

10. The LEAP voltage probe is merely a pair of wires with a phone-cord link at one end and separate alligator clips at the other. The LEAP System box, into which probes are plugged, measures the potential differences as a part of its analog-to-digital signal conversion.

Sample Data and Analysis

Best choice for cell: $Mg/Mg^{2+}//Cu^{2+}/Cu$	
E^0: 2.72 V	
E^0 for two cells in series: 5.44 V	
Balanced redox equation: $Mg(s) + Cu^{2+}(aq) \longrightarrow Mg^{2+}(aq) + Cu(s)$	
Measured voltage: 5.3 V	
Diagram of final combination of cells:	

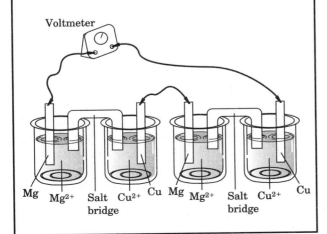

Mg Mg^{2+} Salt bridge Cu^{2+} Cu   Mg Mg^{2+} Salt bridge Cu^{2+} Cu

Voltmeter

Student Orientation

Techniques to Demonstrate

Discuss any errors in technique that occurred during the Exploration. Remind students about reversing the leads if the needle on the voltmeter moves in the wrong direction or if the LEAP System voltage probe shows a zero or negative value. Show students how to link cells in series and how to use a voltmeter or voltage probe to measure voltage across the combination of cells.

Pre-Lab Discussion

Students may become disheartened when they discover that there is not a single cell that will yield a potential of 5.0 V with the materials available. They may need prompting to help them realize that they must evaluate what combinations of cells in series will yield this potential. Be certain students realize that although the cells are connected with a wire, this does not put the copper from one cell in danger of reacting with the magnesium in another cell.

Tips for Evaluating the Pre-Lab Requirements

Above all else, before students enter the lab, make certain that they have made specific choices of which cells and combinations they will test. It is probably not feasible to test more than two or three possible configurations in one lab period. Although students could stumble on the correct combination through trial and error, require them to make choices. No student should be allowed to begin lab work until they have calculated all of the standard potentials for all of the possible cells.

Proposed Procedure

Use two beakers and a salt bridge or one beaker and a porous cup to construct a voltaic cell with copper and magnesium. Immerse each metal in 0.5 M solutions of their ions in beakers or in a beaker and porous cup. Make a second cell similar to the first cell. Using wires with alligator clips, connect the copper strip of the first cell to the magnesium strip of the second cell. Connect either a voltmeter or LEAP System with voltage probe to the magnesium strip of the first cell and the copper strip of the second cell. Change terminals as necessary to give a positive reading or a needle movement to the right. Measure and record the voltage. This procedure may be repeated for additional cells.

Post-Lab

Disposal

Set out 10 labeled disposal containers, one for each metal and solution. The metals should be saved and reused. The solutions should also be saved and reused. The solutions can be evaporated, and the dry compounds can be saved for reuse. Even with the possibility of small amounts of contamination, they are acceptable for use the next time students perform this Exploration. **Do not pour these solutions down the drain. Do not place the metals, solutions, or dry compounds in the trash can.** Although Mg, Sn, Al, $SnCl_2$, and $MgSO_4$ are not considered hazardous wastes, to avoid confusion and disposal accidents, it is best to save and reuse all of these materials.

Additional Notes

Holt Chemistry
Visualizing Matter

LABORATORY EXPERIMENTS

HOLT, RINEHART AND WINSTON
Harcourt Brace & Company

Austin • New York • Orlando • Atlanta • San Francisco • Boston • Dallas • Toronto • London

Lab Authors

Dave Jaeger
Chemistry Teacher
Will C. Wood High School
Vacaville, CA

Suzanne Weisker
Science Teacher and Department Chair
Will C. Wood High School
Vacaville, CA

Safety and Disposal Reviewer

Jay A. Young, Ph.D.
Chemical Safety Consultant
Silver Spring, MD

Teacher Reviewers

Cheryl Epperson
Chemistry Teacher
Flour Bluff High School
Corpus Christi, TX

Marilyn Lawson
Chemistry Teacher/Science Department Chair
Gregory Portland High School
Portland, TX

Cover: (coal) Telegraph Colour Library/FPG International Corp., (diamond structure, orbital) Foca, (diamonds) © 1991 Martin Rogers/FPG International Corp., (full cover wrap) © Goavec Photography/The Image Bank
Illustrations: Progressive Information Technologies

Printed in the United States of America

ISBN 0-03-095284-0

123456 129 98979695

Table of Contents

Working in the World of a Chemist

Meeting today's challenges

Even though you have already taken science classes with lab work, you will find the two types of laboratory experiments in this book organized differently from those you have done before. The first type of lab is called an *Exploration,* and it helps you gain skills in lab techniques that you will use to solve a real problem presented in the second type of lab, which is called an *Investigation.* The *Exploration* serves as a *Technique Builder,* and the *Investigation* is presented as an exercise in *Problem Solving.*

Both types of labs refer to you as an employee of a professional company, and your teacher has the role of supervisor. Lab situations are given for real-life circumstances to show how chemistry fits into the world outside of the classroom. This will give you valuable practice with real-world skills, such as creating a plan with available resources, creating a budget, and writing business letters.

As you work on these labs, you will better understand how the concepts you studied in the chapters are used by chemists to solve problems that affect life for everyone.

Explorations

The *Explorations* provide step-by-step procedures for you to follow, encouraging you to make careful observations and interpretations as you progress through the lab session. There is a series of items and questions at the end of each one that are designed to help you make sense of your data and observations and relate them to the key chemistry concepts being studied.

What you should do before an Exploration

Preparation will help you work safely and efficiently. The evening before a lab, be sure to do the following.

- **Read the lab procedure** to make sure you understand what you will do.

- **Read the safety information** that begins on page vii, as well as that provided in the lab procedure.

- **Write down any questions** you have in your lab notebook so that you can ask about them before the lab begins.

- **Prepare all necessary data tables** so that you will be able to concentrate on your work when you are in the lab.

What you should do after an Exploration

Most teachers require a lab report as a way of making sure that you understood what you were doing. Your teacher will give you specific details about how to organize the lab report, but most lab reports will include the following. A sample lab report you can use as a model is shown in your textbook on page 645.

- **title** for the lab

- **summary paragraph(s)** describing the purpose and procedure

- **data tables and observations** that are organized and comprehensive

- **worked-out calculations** with proper units

- **answers** that are boxed, circled, or highlighted, for items in the *Analysis and Interpretation, Conclusions,* and *Extensions* sections

Investigations

The *Investigations* may seem quite different to you because they do not provide step-by-step instructions. The *Investigations* require you to develop your own procedure to solve a problem presented to your company by a client. As a part of the research team doing work for the client, you must decide how much money to spend on the project and what equipment to use. Although this may seem very difficult, the *Investigations* contain a number of clues about what to do to be successful in *Problem Solving.*

What you should do before an Investigation

Before you will be allowed to work on the lab, you must turn in a preliminary report. Usually, you must describe in detail the procedure you plan to use, provide complete data tables for the data and observations you will collect, and list exactly what equipment you will need and the costs. Only after your teacher, acting as your supervisor, approves your plans are you allowed to proceed. A sample preliminary report is shown on page 646 of the textbook. Before you begin writing a preliminary report, follow the steps on the next page.

- **Read the *Investigation* thoroughly,** searching for clues.

- **Jot down notes** in your lab notebook as you find clues.

- **Consider what you must measure or observe** to solve the problem.

- **Think about *Explorations*** you have done that used a similar technique or reaction.

- **Imagine working through a procedure,** keeping track of each step and what equipment you will need.

- **Carefully consider** whether your approach is the best, most efficient one.

What you should do after an Investigation

After you finish, organize a report of your data as described in the memorandum contained in the *Investigation*. This report is usually in the form of a one- or two-page letter to the client. (Your teacher may have additional requirements for your report.) Carefully consider how to convey the information that the client needs to know. In some cases, a graph or diagram can communicate information better than words can. As a part of your report, you must include an invoice for the client that explains how much they owe and how much you charged for each part of the procedure. Remember to include the cost of your labor during the analysis.

Tips for Success in the Lab

Whether you are performing an *Exploration* or an *Investigation,* you can do the following to help ensure success.

- **Read each lab twice** before you prepare the data tables.

- **Make sure you understand** everything that will happen in the lab.

- **Read the safety precautions** in the lab and on pages vii–x.

- **Prepare the data tables** before you enter the lab.

- **Record all data and observations immediately** in your lab notebook.

- **Use appropriate units** when recording data.

- **Keep your lab desk organized** and free of clutter.

Lab Report Grades

Lab report requirements and grading criteria vary from teacher to teacher. A good lab report usually has these characteristics.

Explorations

- No significant errors in technique occurred during the lab.

- Data and observations are written accurately, descriptively, and completely.

- There are no errors in calculations.

- Good reasoning and logic are shown.

- Students seem to understand what the lab was about.

- Items from the *Analysis and Interpretation* and *Conclusions* sections are answered clearly, concisely, and accurately, with correct units given.

- Any graphs are drawn and labeled accurately and neatly.

Your teacher may have additional requirements.

Investigations

- Preliminary report shows careful and thorough planning with good reasoning and logic.

- Proposed procedure is appropriate and safe.

- Proposed data tables explicitly detail all measurements that must be made to solve the problem.

- Proposed list of equipment includes everything necessary and nothing unnecessary.

- No significant errors in technique occurred during the lab.

- Report or letter is concise, clear, and well organized.

- Data and analysis calculations are accurate, with correct units and no errors in the calculations.

- Percent error for quantitative answers is less than 15%.

- Students seem to understand what the lab was about.

- Any graphs are drawn and labeled accurately and neatly.

- Costs for materials actually used are kept under the budgeted figure.

Your teacher may have additional requirements.

Safety in the Chemistry Laboratory

Chemicals are not toys

Any chemical can be dangerous if it is misused. Always follow the instructions for the experiment. Pay close attention to the safety notes. Do not do anything differently unless told to do so by your teacher.

Chemicals, even water, can cause harm. The trick is to know how to use chemicals correctly so that they will not cause harm. If you follow the rules stated in the following pages, pay attention to your teachers directions, and follow the cautions on chemical labels and the experiments, then you will be using chemicals correctly.

These safety rules always apply in the lab

1. **Always wear a lab apron and safety goggles.**
   Even if you aren't working on an experiment at the time, laboratories contain chemicals that can damage your clothing. Keep the strings of your lab apron tied. More importantly, some chemicals can cause eye damage, even blindness. If your safety goggles are uncomfortable or get clouded up, you are not the first person to have these problems. Ask your teacher for help. Try lengthening the strap a bit, washing the goggles with soap and warm water, or using an anti-fog spray.

2. **No contact lenses are allowed in the lab.**
   Even while wearing safety goggles, chemicals could get between contact lenses and your eyes and cause irreparable eye damage. If your doctor requires that you wear contact lenses instead of glasses, then you should wear eye-cup safety goggles in the lab. Ask your doctor or your teacher how to use this very important and special eye protection.

3. **NEVER WORK ALONE IN THE LABORATORY.**
   You should always do lab work *only* under the supervision of your teacher.

4. **Wear the right clothing for lab work.**
   Necklaces, neckties, dangling jewelry, long hair, and loose clothing can cause you to knock things over or catch items on fire. Tuck in neckties or take them off. Do not wear a necklace or other dangling jewelry, including hanging earrings. It isn't necessary, but it might be a good idea to remove your wristwatch so that it is not damaged by a chemical splash.

   Pull back long hair, and tie it in place. Nylon and polyester fabrics burn and melt more readily than cotton, so wear cotton clothing if you can. It's best to wear fitted garments, but if your clothing is loose or baggy, tuck it in or tie it back so that it does not get in the way or catch on fire.

   Wear shoes that will protect your feet from chemical spills—no open-toed shoes or sandals and no shoes with woven leather straps. Shoes made of solid leather or a polymer are much better than shoes made of cloth. Also, wear pants, not shorts or skirts.

5. **Only books and notebooks needed for the experiment should be in the lab.**
   Do not bring other textbooks, purses, bookbags, backpacks, or other items into the lab; keep these things in your desk or locker.

6. **Read the entire experiment before entering the lab.**
   Memorize the safety precautions. Be familiar with the instructions for the experiment. Only materials and equipment authorized by your teacher should be used. When you do the lab work, follow the instructions and the safety precautions described in the directions for the experiment.

7. **Read chemical labels.**
   Follow the instructions and safety precautions stated on the labels.

8. **Walk with care in the lab.**
   Sometimes you will have to carry chemicals from the supply station to your lab station. Avoid bumping into other students and spilling the chemicals. Stay at your lab station at other times.

9. **Food, beverages, chewing gum, cosmetics, and smoking are NEVER allowed in the lab.**
   (You should already know this.)

10. **NEVER taste chemicals or touch them with your bare hands.**
Also, keep your hands away from your face and mouth while working, even if you are wearing gloves.

11. **Use a sparker to light a Bunsen burner.**
Do not use matches. Be sure that all gas valves are turned off and that all hot plates are turned off *and* unplugged when you leave the lab.

12. **Be careful with hot plates, Bunsen burners, and other heat sources.**
Keep your body and clothing away from flames. Do not touch a hot plate after it has just been turned off. It is probably hotter than you think. The same is true of glassware, crucibles, and other things after you remove them from a hot plate, drying oven, or the flame of a Bunsen burner.

13. **Do not use electrical equipment with frayed or twisted cords or wires.**

14. **Be sure your hands are dry before using electrical equipment.**
Before plugging an electrical cord into a socket, be sure the electrical equipment is turned off. When you are finished with it, turn it off. Before you leave the lab, unplug it, but be sure to turn it off FIRST.

15. **Do not let electrical cords dangle from work stations; dangling cords can cause tripping or electrical shocks.**
The area under and around electrical equipment should be dry; cords should not lie in puddles of spilled liquid.

16. **Know fire drill procedures and the locations of exits.**

17. **Know the location and operation of safety showers and eyewash stations.**

18. **If your clothes catch on fire, walk to the safety shower, stand under it, and turn it on.**

19. **If you get a chemical in your eyes, walk immediately to the eyewash station, turn it on, and lower your head so that your eyes are in the running water.**
Hold your eyelids open with your thumbs and fingers, and roll your eyeballs around. You have to flush your eyes continuously for at least 15 min. Call your teacher while you are doing this.

20. **If you have a spill on the floor or lab bench, call your teacher rather than trying to clean it up by yourself.**
Your teacher will tell you if it is OK for you to do the cleanup; if it is not, your teacher will know how the spill should be cleaned up safely.

21. **If you spill a chemical on your skin, wash it off under the sink faucet, and call your teacher.**
If you spill a solid chemical on your clothing, brush it off carefully so that you do not scatter it on anyone else, and call your teacher. If you get a liquid on your clothing, wash it off right away if you can get it under the sink faucet, and call your teacher. If the spill is on your pants or somewhere else that will not fit under the sink faucet, use the safety shower. Remove the pants or other affected clothing while under the shower, and call your teacher. (It may be temporarily embarrassing to remove your pants or other clothing in front of your class, but failing to flush that chemical off your skin could cause permanent damage.)

22. **The best way to prevent an accident is to stop it before it happens.**
If you have a close call, tell your teacher so that you and your teacher can find a way to prevent it from happening again. Otherwise, the next time, it could be a harmful accident instead of just a close call.

23. **All accidents should be reported to your teacher, no matter how minor.**
Also, if you get a headache, feel sick to your stomach, or feel dizzy, tell your teacher immediately.

24. **For all chemicals, take only what you need.**
On the other hand, if you do happen to take too much and have some left over, DO NOT put it back in the bottle. If somebody accidentally puts a chemical into the wrong bottle, the next person to use it will have a contaminated sample. Ask your teacher what to do with any leftover chemicals.

25. **NEVER take any chemicals out of the lab.**
(This is another one that you should already know. You probably know the remaining rules also, but read them anyway.)

26. Horseplay and fooling around in the lab are very dangerous.
NEVER be a clown in the laboratory.

27. Keep your work area clean and tidy.
After your work is done, clean your work area and all equipment.

28. Always wash your hands with soap and water before you leave the lab.

29. Whether or not the lab instructions remind you, ALL of these rules APPLY ALL OF THE TIME.

Safety Symbols

To highlight specific types of precautions, the following symbols are used in the *Explorations* and *Investigations*. **Remember that no matter what safety symbols are used, all of the 29 safety rules described previously should be followed at all times.**

- Wear laboratory aprons in the laboratory. Keep the apron strings tied so that they do not dangle.
- Wear safety goggles in the laboratory at all times. Know how to use the eyewash station. (Rules **1, 17,** and **19** apply. Which rule says not to wear contact lenses? Do regular eyeglasses provide enough protection?)

- Never taste, eat, or swallow any chemicals in the laboratory. Do not eat or drink any food from laboratory containers. Beakers are not cups, and evaporating dishes are not bowls. (Rules **9** and **10** apply.)
- Never return unused chemicals to the original container. (Which rule applies?)
- Some chemicals are harmful to our environment. You can help protect the environment by following the instructions for proper disposal.
- It helps to label the beakers and test tubes containing chemicals. (This is not another new rule, just a good idea.)

- Never transfer substances by sucking on a pipette or straw; use a suction bulb. (This one is another new rule.)
- Never place glassware, containers of chemicals, or anything else near the edges of a lab bench or table. (This is another new rule.)

Here is a question: Are there any rules other than **9** and **10** that apply in the lab when you see this safety symbol?

- If a chemical gets on your skin or clothing or in your eyes, rinse it immediately, and alert your teacher. (Rules **19** and **21** apply.)
- If a chemical is spilled on the floor or lab bench, tell your teacher, but do not clean it up yourself unless your teacher says it is OK to do so. (Rule **20** applies.)

Here is another question: What rules other than **19, 20,** and **21** apply in the lab when you see this safety symbol?

- When heating a chemical in a test tube, always point the open end of the test tube away from yourself and other people. (This is another new rule.)

WARNING! From here to the end of this safety section, it is up to you to look at the list of rules and identify whether a specific rule applies or if the rule presented is a new rule.

- Tie back long hair, and confine loose clothing. (Rule **?** applies.)
- Never reach across an open flame. (Rule **?** applies.)
- Use proper procedures when lighting Bunsen burners. Turn off hot plates, Bunsen burners, and other heat sources when not in use. (Rule **?** applies.)
- Heat flasks and beakers on a ring stand with wire gauze between the glass and the flame. (Rule **?** applies.)

- Use tongs when heating containers. Never hold or touch containers with your hands while heating them. Always allow heated materials to cool before handling them. (Rule **?** applies.)

- Turn off gas valves when not in use. (Rule **?** applies.)

- Use flammable liquids only in small amounts. (Rule **?** applies.)

- When working with flammable liquids, be sure that no one else in the lab is using a lit Bunsen burner or plans to use one. Make sure there are no other heat sources present. (Rule **?** applies.)

  What other rules should you follow in the lab when you see this safety symbol?

- Check the condition of glassware before and after using it. Inform your teacher of any broken, chipped, or cracked glassware because it should not be used. (Rule **?** applies.)

- Do not pick up broken glass with your bare hands. Place broken glass in a specially designated disposal container. (Rule **?** applies.)

- Never force glass tubing into rubber tubing, rubber stoppers, or wooden corks. To protect your hands, wear heavy cloth gloves or wrap toweling around the glass and the tubing, stopper, or cork, and gently push in the glass. (Rule **?** applies.)

- Do not inhale fumes directly. When instructed to smell a substance, use your hand, wave the fumes toward your nose, and inhale gently. (Some people say "waft the fumes.")

- Keep your hands away from your face and mouth. (Rule **?** applies.)

- Always wash your hands before leaving the laboratory. (Rule **?** applies.)

Finally, if you are wondering how to answer the questions that asked what additional rules apply to the safety symbols, here is the correct answer. **Any time you see any of the safety symbols you should remember that all 29 of the numbered laboratory rules always apply.**

1-1 Conservation of Mass

Technique Builder

Objectives

Demonstrate proficiency in detecting a chemical reaction.

Measure masses of reactants and products with a laboratory balance.

Design chemical experiments.

Relate observations of a chemical reaction to the law of conservation of mass.

Situation

Your company has received a contract to develop a series of lab exhibits to be used at a *Fun Science* display that will be traveling to shopping malls across the country. The first exhibit requires participants to observe a chemical reaction between two household substances, vinegar and baking soda, and then design a way to contain the gas given off by the reaction so that the mass can be measured to determine if it is conserved.

Background

The law of conservation of mass states that matter is neither created nor destroyed during a chemical reaction and that the mass of a system should therefore remain constant during any chemical process. This means that during any reaction the sum of the masses of the products of a reaction must be the same as the sum of the masses of the reactants. For example, imagine using a spark to ignite a match in a glass jar which is so tightly sealed that no matter can enter or escape. As the match burns, flammable substances from the match and some oxygen from the air are transformed into water, carbon dioxide, and other substances. Measurements taken before and after the match burned would show that the total mass remains the same.

Although this concept may seem obvious, it was not always so. When early scientists first began exploring chemical changes, they frequently did not consider the effects of the air and other gases. Similarly, the equipment originally used often did not take into account the possibility that a gas could be a product or a reactant.

Problem

You will need to perform the following steps for this reaction to develop the exhibit.

- Accurately measure the mass of reactants and products in a controlled chemical reaction in an open container.
- Design a method to perform the same chemical reaction in a closed container so that no products escape.
- Accurately measure the mass of reactants and products in a controlled chemical reaction in a closed container.
- Compare your results with those of other groups.
- Provide instructions for the exhibit.

Safety

Always wear goggles and a lab apron to provide protection for your eyes and clothing. If you get a chemical in your eyes, immediately flush it out at the eyewash station while calling to your teacher. Know the locations of the emergency lab shower and eyewash and how to use them.

Do not touch any chemicals. If you get a chemical on your skin or clothing, wash it off at the sink while calling to your teacher. Make sure you carefully read the labels and follow the directions on all containers of chemicals that you use. Do not taste any chemicals or items used in the laboratory. Never return leftovers to their original containers; take only small amounts to avoid wasting supplies.

Always clean up the lab and all equipment after use, and dispose of substances according to proper disposal methods. Wash your hands thoroughly before you leave the lab after all lab work is finished.

Materials

- Baking soda ($NaHCO_3$)
- Vinegar (acetic acid, CH_3COOH, solution)
- 400 mL beaker
- 100 mL graduated cylinder
- Balance
- Clear plastic cups (or 150 mL beakers), 2
- Large plastic bag
- Twist ties, 2
- Weighing papers, 2

Preparation

1. Organizing Data
Make a data table in your lab notebook with four columns and three rows. Label the boxes in the first row of the second, third, and fourth columns *Initial mass (g), Final mass (g),* and *Change in mass (g).* Label the boxes in the second and third rows of the first column *Part 1* and *Part 2.* Below the table, leave room for observations about the reaction in Parts 1 and 2.

Technique

Part 1

2. Place a piece of weighing paper on the laboratory balance. Place about 4–5 g of baking soda on the weighing paper. When you have that much, carefully transfer the baking soda to a plastic cup or beaker. ***Never put any chemical directly on the balance pan of a laboratory balance. Never return any excess chemical to its original container because the risk that you might accidentally put the wrong chemical into the container is too great.***

3. Using the 100 mL graduated cylinder, measure about 100 mL of vinegar. Remember to read the volume at the bottom of the meniscus, the curved surface that the water forms where it meets the air. Pour the vinegar into the second plastic cup (or beaker).

4. Place both cups on the balance pan of a laboratory balance, and determine the initial total mass of the system to the nearest 0.01 g. Record this mass in your data table. (If the balance pan is too small to hold both cups, measure the mass of one, and then the other. Add these masses to find the total mass.)

5. Take the cups off the balance. Carefully pour the vinegar into the cup that contains the baking soda. Pour gently, along the inside wall of the baking soda cup, so splashing is kept to a minimum. It may help the reaction to reach completion sooner if you gently swirl the cup to make sure the reactants are well-mixed. Record your observations in your lab note-book. ***Add only a small amount of vinegar at a time to avoid a strong reaction that causes splattering and the loss of part of the reactants or products.***

6. When the reaction has finished, place both cups back on the balance to determine the final mass of the system to the near-est 0.01 g. Record the final mass in your data table. Calculate any change in mass.

Part 2

7. Examine the plastic bag and the twist ties. Develop a proce-dure that will test the law of conservation of mass more accu-rately than does the procedure in Part 1. When you think you know what to do, ask your teacher to approve your plan. (Hint: it will be easier to measure the mass if you rest the plastic bag in the 400 mL beaker. Do not use more than 50 mL of vinegar.)

8. After your teacher approves your plan, implement it using the same materials and quantities as in Part 1. Use the plastic bag and the twist ties instead of the cups from Part 1. Re-member to stop and record your data in your data table after each step.

9. If you were successful in step **8** and your results reflect that mass was conserved, proceed. If not, find a lab group that was successful, and find out what they did and why. Test their procedure to determine whether the results of the other group are reproducible.

Cleanup and Disposal

10. Place each product or reactant in the separate disposal con-tainers provided by your teacher. Clean up your equipment and lab station. Thoroughly wash your hands after complet-ing the lab work and cleanup.

Analysis and Interpretation

1. **Organizing Conclusions**
   Describe all evidence indicating that a chemical change occurred in this experiment.

Conclusions

2. **Evaluating Methods**
   Was the law of conservation of mass really violated in Part 1 as it appeared? Explain your answer.

3. **Evaluating Methods**
   Explain why the results for Part 2 were different from those for Part 1. What is the system in Part 1? What is the system in Part 2? How do they differ?

4. **Organizing Ideas**
   How can this lab be used to prepare the *Fun Science* exhibits? Explain your proposal for packaging the contents, and write instructions for the participants who will visit the exhibit.

5. Evaluating Methods

Explain how the steps of the scientific method were used in this procedure. (Hint: see the discussion of the scientific method in Section 1-2 for ideas.)

6. Applying Ideas

You can tell from the title of this Exploration that one of the themes from Chapter 1 that applies to this procedure is the theme of conservation. Pick another theme from Section 1-3, and explain how it relates to what happened during the procedure.

Extensions

1. Applying Conclusions

When a log burns, the resulting ash obviously has less mass than the log did. Explain whether this loss of mass violates the law of conservation of mass.

2. Research and Communications

The law of conservation of mass was first stated by Antoine Lavoisier in 1789. Prepare a report about Lavoisier's work and the textbook that he published.

3. Research and Communications

Just as mass is conserved in chemical reactions, so is energy. Find out several ways in which energy is converted and transferred when an automobile is driven. Present your findings as a poster to share with your class.

4. Research and Communications

The reaction that takes place in this Exploration can be summarized as follows.

$$NaHCO_3 + CH_3COOH \longrightarrow Na\ CH_3COO + H_2O + CO_2$$

| sodium bicarbonate | acetic acid (dissolved) | sodium acetate (dissolved) | water | carbon dioxide gas |

Look up each substance in a chemical handbook and prepare a fact sheet that includes information such as melting point, boiling point, and density.

5. Research and Communications

Using a chemical handbook or other reference, compile a list of the hazards associated with each of the reactants and products. Prepare a presentation for the class about whether or not these chemicals are too dangerous to be used in a shopping mall. Thoroughly explain your reasoning.

2-1 Separation of Mixtures

Objectives

Identify chemical and physical properties of substances.

Relate knowledge of properties to the task of separating mixed items.

Identify as many methods for separating the items as possible.

Separate the components of a mixture.

Analyze success of methods for purifying mixtures.

Design and implement your own procedure.

Situation

You work for a company that has joined the adopt-a-school program. The Springfield Consolidated School District is holding a Science Olympics, and your company has volunteered to prepare challenge packages containing a mixture of sand, salt, iron filings, and poppy seeds. The participants in the Science Olympics will be challenged to recover each of the components of the mixture as separate items. Your supervisor has asked you to try the challenge so that you can develop guidelines for evaluating the contestants' procedures.

Background

As was discussed in Chapter 2, a mixture is a combination of two or more kinds of matter. The different kinds of matter in a mixture can be separated by physical means because each component in a mixture retains its own composition and properties, as discussed in Section 2-4. Some methods will work for some components, but not for others. Another issue to keep in mind is how easy it will be to reclaim the different components of a mixture. For example, although paper chromatography can indicate the number and colors of components in a mixture, it can be difficult to use as a method for *recovering* the components.

Problem

In order to prepare guidelines for evaluating the task, you must do the following.
- Identify as many physical and chemical properties of the substances in the mixture as you can.
- Figure out how to use these properties to identify as many different methods of separation as you can.
- Evaluate the methods to determine which is the best.
- Perform this method, and record the time it takes.

Safety

Always wear goggles and a lab apron to provide protection for your eyes and clothing. If you get a chemical in your eyes, immediately flush it out at the eyewash station while calling to your teacher. Know the locations of the emergency lab shower and eyewash and how to use them.

Do not touch any chemicals. If you get a chemical on your skin or clothing, wash it off at the sink while calling to your teacher. Make sure you carefully read the labels and follow the directions on all containers of chemicals that you use. Do not taste any chemicals or items used in the laboratory. Never return leftovers to their original containers; take only small amounts to avoid wasting supplies.

Always clean up the lab and all equipment after use, and dispose of substances according to proper disposal methods. Wash your hands thoroughly before you leave the lab after all lab work is finished.

Preparation

Organizing Methods

1. Before you begin, you will need to develop a plan for separating the components of the mixture. Start by trying to determine which properties of a component in the mixture are not shared by most of the other components. When you think you know what to do, **write down the entire plan in your lab notebook before you proceed.** Be sure to estimate how long each step will take so that you can plan your time in lab more effectively.

2. If any part of your procedure involves a long period of waiting, try to reach that point before the end of the class period. Write your name on the side of your petri dish, and use it as a holding bin for your setup. Ask your teacher where you can leave it until you can complete your procedure on the next lab day.

Technique

3. Obtain a sample of the mixture, a petri dish, a microfunnel, and an 8-well microchemistry strip. Using the microfunnel, place a small part of your sample in the first well of the 8-well microchemistry strip so that you will be able to compare your separated components to the original mixture. Place a small piece of tape over the opening of the well so that the contents will not spill out.

4. Record the time when you begin your work in your lab notebook.

5. Using any or all of the items listed in the materials list, implement the procedure you wrote in your lab notebook to separate and recover all four components of the rest of the mixture. Make as many observations as possible at each step, and record exactly what you do in your lab notebook.

Materials

- Distilled water
- Sample of mixture
- 8-well micro-chemistry strip
- Aluminum foil
- Cellophane
- Cotton balls
- Filter paper
- Forceps
- Glass funnel
- Magnets
- Microfunnel
- Paper clips
- Paper towels
- Petri dish
- Pipets
- Plastic forks
- Plastic spoons
- Plastic straws
- Rubber stoppers
- Tape
- Test-tube holder
- Test-tube rack
- Test tubes
- Tissue paper
- Wood splints

6. When you have separated the first component from the mixture, place a small amount of it in the third well of the 8-well microchemistry strip, leaving the second well empty between the mixture in the first well and the purified component in the third well. Place a small piece of tape over the opening of the well.

7. As each component is separated, place small amounts in the fourth through sixth wells of the 8-well microchemistry strip. Place a small piece of tape over the opening of each well after it is filled, so that the contents do not spill out.

8. When all of the components have been separated, ask your teacher to inspect the samples and approve your work. If it is approved, record how long it has taken in your lab notebook. Write this on a small piece of paper. Place the paper and the 8-well microchemistry strip in your petri dish, and put them in the location indicated by your teacher.

9. Examine the strips of other lab groups. Record observations about the purity of each lab group's samples of salt, sand, iron filings, and poppy seeds. Record observations about the amount of time each lab group took.

Cleanup and Disposal

10. Your teacher should have placed several disposal containers at the front of the room. After the appropriate wells have been filled with small samples of the recovered components, place the remainder of each recovered component in the appropriate disposal containers. Clean up your equipment and lab station. Thoroughly wash your hands after completing the lab work and cleanup.

Analysis and Interpretation

1. Analyzing Results
Write a paragraph summarizing your procedure. In your estimation, rate the recovery of each component on a scale of 1–10. Justify the estimations of your success based on both the purity of your components and the time it took for you to do the job.

Conclusions

2. Evaluating Methods
What made you decide to do your procedural steps in the order that you did them? Would any order have worked?

3. Analyzing Methods
If you were able to do the lab over again, what specific things would you do differently?

4. Applying Ideas
Name any materials or tools that were not available that might have made the separation of the substances easier.

5. Analyzing Conclusions
For each of the four components, describe a specific physical property that enabled you to separate it from the rest of the mixture.

6. Evaluating Methods

Discuss the relationship you expect to find between the speed of the process and the purity of the components recovered. Which do you think is more important? Explain your answer.

7. Evaluating Methods

Create a plan for scoring the results of students participating in the Science Olympics. Apply this plan to the observations you made about the other lab groups' results, and provide a list of scores for your class.

Extensions

1. Applying Information

What methods could you use to determine the purity of each of your recovered components?

2. Applying Ideas

How would you separate each of the following two-part mixtures?

a. lead filings and iron filings
b. sand and gravel
c. sand and finely ground plastic foam
d. salt and sugar
e. alcohol and water
f. nitrogen and oxygen

3. Research and Communication

Find out about methods used to produce large quantities of products from coal or oil. Prepare a chart or poster to explain the different stages of separation of these products.

4. Research and Communication

Investigate the ways in which drinking water is purified in your community. Prepare a presentation to your class about how these methods work and why they must be used.

Problem Solving

2-1

January 23, 1995

Marissa Bellinghausen
Director of Investigations
CheMystery Labs, Inc.
52 Fulton Street
Springfield, VA 22150

Dear Ms. Bellinghausen:

A tanker truck carrying approximately 3.8×10^4 L of waste from a food-processing plant overturned along an embankment at the edge of Zavala Lake. The tanker carried no manifest, but its contents were apparently ground coffee, garlic powder, and mineral oil. Although these spilled products are not believed harmful, they make the water look and smell unpleasant. The lake is a favorite recreational spot used by many people for swimming, sailing, and water-skiing, so this is a problem that needs to be cleaned up.

Preliminary sample analysis of the lake water suggests that sand and/or charcoal may be adequate for cleaning the water. We are requesting that your company analyze the sample to evaluate the effectiveness of sand and charcoal as treatment mediums and to suggest detailed procedures for cleanup. A sample of the contaminated water will arrive under separate cover, as will a contract for $70 000.

Sincerely,

Anita Camacho

Anita Camacho
Director
Department of Health Services
Verde County

Memorandum

Date: May 10, 1995
To: Angelo Portelli
From: Marissa Bellinghausen

CheMystery Labs, Inc.
52 Fulton Street
Springfield, VA 22150

If we do a good job planning the purification, we will probably be hired to help implement it, which will provide steady revenue for the company. This is crucial to helping us stay in business.

I think you should start by letting the sample settle. Then, I recommend some filtering steps. Because the final cleanup will involve huge volumes of water, I don't think using a filtering system with filter paper will be practical. Instead, consider a system in which water is pumped through several tubes of material. To model this system, pour the water into a long-stemmed microfunnel with about 15 g of sand and another long-stemmed microfunnel with about 8 g of charcoal. (Hint: to keep these substances in the microfunnel, place a wad of tissue paper or cotton in the bottom first.)

Because we don't have a large budget, I'll need to look over your plan before you proceed. Prepare a proposal that includes the following.

- detailed one-page plan for the procedure
- illustrations or descriptions of your filter bed constructions
- listing of the properties of the mixture's components that you will use to separate them
- list of all necessary materials and equipment, along with individual and total costs for the accounting department

When you are finished with your analysis, prepare a report to Anita Camacho in the form of a two-page letter that includes the following.

- summary of your plan to clean up the lake that relates the properties of the components to your plan and reflects the best use of filters.
- justification of your proposed filtering setup
- discussion of the benefits of sand as a water purifier and the benefits of charcoal as a purifier
- identify the sand filter construction that minimizes filtration time while maximizing water purity and explain why it works best
- data on color, clarity, odor, presence of oils or solids, and volume for each step of your procedure
- percentage of initial volume of H_2O remaining after treatment
- discussion of the likelihood that impurities remain in the water
- detailed invoice for all costs

Required Precautions

 Always wear goggles and a lab apron to provide protection for your eyes and clothing.

 Do not touch any chemicals. If you get a chemical on your skin or clothing, wash it off at the sink while calling to your teacher.

 Always wash your hands thoroughly when finished.

MATERIALS for Verde County

REQUIRED ITEMS
(You must include all of these in your budget.)

Lab space / fume hood / utilities	15 000 / day
Standard liquid disposal fee	20 000 / L of product

REAGENTS AND ADDITIONAL EQUIPMENT
(Include in your budget only what you will need.)

Charcoal	500 / g
Sand (mixture of coarse and fine)	500 / g
25 mL graduated cylinder	1 000
50 mL beaker	1 000
250 mL beaker	1 000
Desiccator	3 000
Glass stirring rod	1 000
Long-stemmed microfunnel	1 000
Pipet bulb without stem	1 000
Pipet (thin-stemmed)	1 000
Plastic spoon	500
Plastic tray	1 000
Ring stand / ring	2 000
Scissors	500
Six test tubes / holder / rack	2 000
Stopwatch	5 000
Strainer	1 000
Tissue paper	500 / piece
Wash bottle	500
Watch glass	1 000
Wire	500
Wooden splint	500

* No refunds on returned chemicals or unused equipment.

FINES

OSHA safety violation	2 000 / incident

Spill Procedures / Waste Disposal Methods

- To recover the sand, hold the sand-filter funnel upside down over the sand disposal container, squeeze the tip, and roll it between your fingers. Use a micro-tip pipet to squirt water through the opening to flush the sand into the container.
- To recover the charcoal, pinch and roll the tip of the pipet, and then use a small piece of wire to push any that remains into the charcoal disposal container.
  - Oily residue on equipment should be absorbed onto paper towels, which may be placed in the trash can. Then wash and scrub the equipment with soap and water.
  - Purified water may be poured down the sink.
  - Any foul water, oil, or other liquids should be collected in the same disposal container.

References

The theory behind mixtures and their separation is discussed in the textbook on pages 52–56. Your team has used the physical properties of materials to separate mixtures of sand, salt, iron filings, and poppy seeds. To separate mixtures that form layers, squeeze a pipet bulb that has most of its stem cut off to pull the mixture into the bulb. Hold the bulb with the open side down, allow the layers to settle, then gently squeeze to empty each layer, one at a time, into different containers. Coarse sand filters out fewer particles and allows water to flow through faster than fine sand. To separate coarse sand from fine sand, pass it through a strainer held over a plastic tray. To measure volume, calibrate a wooden splint so that each mark on the stick corresponds to the height of 1.0 mL in the test tubes you will be using.

Covalent and Ionic Bonding—
Ceramics Fixative

February 17, 1995

Mr. Reginald Brown
Director of Materials Testing
CheMystery Labs, Inc.
52 Fulton Street
Springfield, VA 22150

Dear Mr. Brown,

In our research for creating a new type of glaze for our ceramic pottery, we have run into a problem. We have tried using a variety of compounds to develop a fixative for the glaze but have failed to find one that meets our criteria.

The compound in the fixative must be transparent when it dissolves and white when it dries so that it will not affect the other colors in the glaze. Next, the compound must have a high melting point for the kiln firing. We would like the fixative to dissolve in water but not in alcohol. Finally, the compound must be electrically conductive when it is dissolved in water. It does not matter whether the fixative is organic or inorganic. We have identified four compounds for study: sucrose, sodium chloride, sodium carbonate, and salicylic acid. You should determine which of these substances are ionic and which are covalent and recommend which will best fit our criteria.

As discussed on the telephone, a contract for $135 000 will arrive under separate cover.

Sincerely,

Kathleen Sylva

Kathleen Sylva
Head Researcher
Ceramic Artisans

Memorandum

Date: February 21, 1995
To: Teresa Rodriguez
From: Reginald Brown

CheMystery Labs, Inc.
52 Fulton Street
Springfield, VA 22150

This company is on the cutting edge of new artistic ceramics materials. If we succeed in doing well on this contract, others could follow.

Because we do not have a kiln or the necessary equipment to do high-temperature studies, I recommend that you gently heat about 0.5 g of each substance in a test tube held in a burner flame. Once the substance melts, remove it from the burner flame. If a substance does not appear to melt after 1.0 min, remove it from the flame. Although we will not be able to measure precise melting points, this will help us determine which substances have the highest melting points.

In order to help ensure quality results, let me review the following before you begin work.
- paragraph summarizing your procedure, including safety precautions you will take
- data table with spaces for all necessary items
- complete list of all materials and equipment, with total and individual costs

When you complete your research, prepare a summary of your results in the form of a two-page letter to Ms. Sylva. Be sure to include the following.
- your recommendation of the best compound
- paragraph discussing your reasons for your recommendation
- paragraph summarizing the procedures used, including a discussion of how the criteria requested were tested
- completed data table
- detailed invoice for materials and services

Required Precautions

 Always wear goggles and a lab apron to provide protection for your eyes and clothing.

 Tie back loose clothing and hair. Do not heat glassware that is broken, chipped, or cracked. Be sure test tubes are heat-resistant before heating. Always use a test-tube holder when heating a tube. Continuously move the tube while heating it to prevent pressure from building up. Never point the opening of a test tube that you are heating toward yourself or others.

 Ethanol is a flammable liquid. Perform the melting point tests first. Then, only after your teacher tells you that all students have finished their melting point tests and all Bunsen burners are turned off and have cooled down, proceed with testing how well the substances dissolve in water and ethanol.

 Do not touch any chemicals. If you get a chemical on your skin or clothing, wash it off at the sink while calling to your teacher.

 Always wash your hands thoroughly when finished.

References

The properties of ionic and covalent bonds are discussed in Chapters 5 and 6. The two types of bonds are compared and contrasted on pages 190–196 of the textbook.

The solubility of 1.0 g of a substance can be determined by shaking it with 5 mL of water or alcohol in a stoppered test tube.

Be sure to rinse the conductivity tester with distilled water between each test.

Use the total mass of the substances you test, including those dissolved in solvents, when calculating disposal costs.

Spill Procedures/ Waste Disposal Methods

- Contact your teacher if a spill occurs.
- Each undissolved sample should go in its own separate disposal container.
- All water solutions should be placed in their own separate container.
- All ethanol solutions should be placed in their own separate container. Do not pour ethanol solutions down the drain.
- Scrub all of the test tubes well with soap, water, and a test-tube brush to remove any residues of the compounds tested.

MATERIALS for Ceramic Artisans

REQUIRED ITEMS
(You must include all of these in your budget.)

Lab space/fume hood/utilities	15 000/day
Standard solid disposal fee	2 000/g
Balance	5 000
Bunsen burner/related equipment	5 000

REAGENTS AND ADDITIONAL EQUIPMENT
(Include in your budget only what you will need.)

Ethanol	1000/mL
Salicylic acid	1 000/g
Sodium carbonate	1 000/g
Sodium chloride	1 000/g
Sucrose	1 000/g
150 mL beaker	1 000
250 mL beaker	1 000
250 mL flask	1 000
100 mL graduated cylinder	1 000
Conductivity probe, LEAP	5 000
Conductivity tester, battery-operated	5 000
Forceps	500
Glass stirring rod	1 000
Rubber stopper (for test tube)	500
Six test tubes/holder/rack	2 000
Spatula	500
Wash bottle	500
Weighing paper	500/piece

* No refunds for returned chemicals or unused equipment.

FINES

OSHA safety violation	2 000/incident

EXPLORATION 6-2
Technique Builder

Viscosity of Liquids

Objectives

Demonstrate proficiency in comparing the viscosity of various liquids under identical test conditions.

Construct a small viscosimeter.

Measure flow time of various single-weight oils.

Measure the mass and volume of the oils to calculate density.

Calculate the relative viscosity of the oils.

Graph experimental data.

Compare viscosities and densities to determine the SAE rating of each oil.

Situation

You have been contacted by an automotive service shop that received a shipment of bulk containers of motor oil. The containers had been shipped by freight train, but several boxcars had leaky roofs. As a result, the labels peeled off the cans. Before the shop uses this oil in cars, they must match up the cans with the types of oil that were listed on the shipping invoice, based on the viscosity and the SAE rating of the oils.

Background

Viscosity is the measurement of a liquid's resistance to flow. Several factors contribute to viscosity. Liquids with high intermolecular forces tend to be very viscous. For example, glycerol has a high viscosity because of its tendency to form many hydrogen bonds. For other molecules, such as oils, the longer the chain length of the molecule, the more viscous they are. The longer chains not only provide greater surface area for intermolecular attractions, but also they can be intertwined more easily. For example, gasoline, which contains molecules that are chains of three to eight carbon atoms, is much less viscous than grease, which usually contains molecules with about 20 to 25 carbon atoms.

The Society of Automotive Engineers rates lubricating oils according to their comparative viscosities. These numerical values, called SAE ratings, range from SAE-10 (low viscosity) to SAE-60 (high viscosity) for oils typically used in combustion engines such as those in automobiles and trucks. The ratings are achieved with an instrument called a viscosimeter, which has a small capillary tube opening. The amount of time for a specific amount of motor oil to flow through the opening is a measure of viscosity. The less viscous oils flow through in a shorter time than the more viscous oils do.

Problem

To match the correct oil sample to its SAE rating, you will need to do the following.
- Make your own viscosimeter from a pipet.
- Measure the relative viscosities of several oils by timing the oil as it flows through your viscosimeter.
- Measure mass and volume of each oil to calculate density.
- From the measurements, infer which labels belong on the containers of oil.

Safety

Always wear goggles and a lab apron to provide protection for your eyes and clothing. If you get a chemical in your eyes, immediately flush it out at the eyewash station while calling to your teacher. Know the locations of the emergency lab shower and eyewash and how to use them.

Do not touch any chemicals. If you get a chemical on your skin or clothing, wash it off at the sink while calling to your teacher. Make sure you carefully read the labels and follow the directions on all containers of chemicals that you use. Do not taste any chemicals or items used in the laboratory. Never return leftovers to their original containers; take only small amounts to avoid wasting supplies.

When you use a Bunsen burner, confine any long hair and loose clothing. Do not heat glassware that is broken, chipped, or cracked. Use tongs or a test-tube holder to handle heated glassware and other equipment because hot glassware does not always look hot. If your clothing catches on fire, WALK to the emergency lab shower, and use it to put out the fire. Because the oil tested in this lab is flammable, it should never be heated directly over a flame. Instead, use a hot-water bath, and never heat it above 60°C.

Pins are sharp; use with care to avoid cutting yourself or others.

Always clean up the lab and all equipment after use, and dispose of substances according to proper disposal methods. Wash your hands thoroughly before you leave the lab after all lab work is finished.

Preparation

1. **Organizing Data**
   Prepare a data table in your lab notebook with 13 columns and 8 rows. Label the boxes in the first row *Sample, Beaker mass (g), Total mass (g), Volume (mL), Trial 1—cool (s), Trial 2—cool (s), Trial 3—cool (s), Trial 1—room temp. (s), Trial 2—room temp. (s), Trial 3—room temp. (s), Trial 1—warm (s), Trial 2—warm (s),* and *Trial 3—warm (s)*. In the first column, label the second through eighth boxes *A, B, C, D, E, F,* and H_2O. Leave three spaces below the table to record the cool temperature, the room temperature, and the warm temperature.

2. With a wax pencil, label each 50 mL beaker-test tube-pipet set with the name of one oil sample (*A, B, C, D, E,* or *F*). Label an additional set H_2O.

3. Place two marks 2.0 cm apart on the side of the bulb of the pipet, as shown in the illustration. The top mark will be the starting point and the lower mark will be the endpoint.

Materials

- Distilled water
- Ice
- Oil samples, 10 mL, 6
- 50 mL beakers, 7
- 400 mL beakers, 2
- 10 mL graduated cylinder
- Pin, straight
- Pipets, thin-stem, 7
- Ruler, metric
- Stopwatch or clock with second hand
- Test-tube holder
- Test-tube rack
- Test tubes, small, 7
- Wax pencil

Bunsen burner option
- Bunsen burner and related equipment
- Ring stand and ring
- Wire gauze with ceramic center

Hot plate option
- Hot plate

Thermometer option
- Thermometer, non-mercury
- Thermometer clamp

LEAP System option
- LEAP System
- LEAP thermistor probe

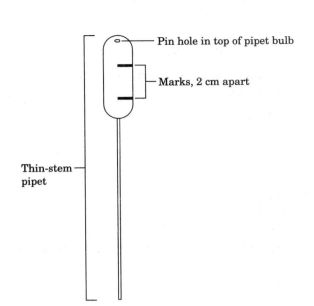

Pin hole in top of pipet bulb

Marks, 2 cm apart

Thin-stem pipet

4. Carefully make a small hole in the top of the bulb of each pipet with the pin, as shown in the illustration. Be sure the hole is well above the marks you made on the side of the pipet bulb. Make the hole the same size for each pipet by putting in the pin the same way for each one. You will control the flow of oil with your finger and this hole.

Technique

5. Measure the masses of all seven 50 mL beakers. Record them in your data table.

6. Pour about 5.0 mL of distilled water into the graduated cylinder. Measure and record the volume to the nearest 0.1 mL, and pour it into the H_2O beaker.

7. Measure and record the mass of the H_2O beaker with water in your data table.

8. Squeeze the H_2O pipet bulb and fill the pipet with distilled water to above the top line. After it is full, place your finger over the pin hole. Place the pipet over the H_2O beaker, lift your finger off the hole, and allow the liquid to flow into the beaker until the meniscus is even with the top line on the pipet bulb. Cover the hole promptly when the water reaches this point. Several practice trials may be necessary.

9. One member of the lab group should hold the pipet with a finger over the pinhole, and the other should use a clock with a second hand or a stopwatch to record precise time intervals. Hold the pipet over the H_2O beaker. When the timer is ready, remove your finger from the pinhole, and allow the liquid to flow into the beaker until it reaches the bottom line on the pipet bulb. Record the time elapsed to the nearest 0.1 s in your data table in the section for room temperature. (If you do not have a stopwatch, measure the time elapsed to the nearest 0.5 s.) It may take several practice trials to master the technique.

10. Repeat steps **6–9** with each oil, using the appropriately labeled pipets and beakers. You should perform several trials for each oil and for water to obtain consistent results.

11. Using one of the 400 mL beakers, make an ice bath. Fill the test tubes to within 1.0 cm of the top with the appropriate oil or distilled water. Cool the samples for 5–8 min so that they are at a temperature between 0°C and 10°C. The key is that all of the samples must be at the same temperature. Measure the temperature of the water sample to the nearest 0.1°C with a thermometer or a LEAP System thermistor probe and record it below your data table.

12. Repeat steps **6–9** with each of the cooled samples. Be sure to use the pipets and 50 mL beakers designated for each oil or distilled water. Record the volume, mass, and time elapsed for each trial in your data table.

13. Using a Bunsen burner or a hot plate and the second 400 mL beaker, prepare a warm-water bath with a temperature between 35°C and 45°C. If you measure the temperature with a thermometer, use a thermometer clamp attached to a ring stand to hold the thermometer in the water.

14. Refill the test tubes to within 1.0 cm of the top with the appropriate oil or distilled water. Place these test tubes into the warm-water bath, and allow the oil and water to warm. Record the temperature of the water sample with a thermometer or a LEAP System with thermistor probe when you remove the samples.

15. Repeat steps **6–9** with the warm samples. Record the volume, mass, and time elapsed for each trial in your data table.

Cleanup and Disposal

16. Your instructor will have set out twelve disposal containers; six for the six types of oil and six for the pipets. **Do not pour oil down the sink. Do not put the oil or oily pipets in the trash can.** The distilled water may be poured down the sink. The test tubes should be washed with a mild detergent and rinsed. Always wash your hands thoroughly after cleaning up the area and equipment.

Analysis and Interpretation

1. Organizing Data
Determine the density of each sample.

2. Organizing Data
Find the average flow time for each sample at each temperature.

3. Analyzing Information
Calculate the relative viscosity of your samples at room temperature by applying the following formula. The values for the absolute viscosity of water are in units of centipoises (cp). A centipoise is equal to 0.01 g/cm·s.

$$\text{relative viscosity}_{oil} = \frac{\text{density}_{oil} \times \text{time elapsed}_{oil} \times \text{viscosity}_{H_2O}}{\text{density}_{H_2O} \times \text{time elapsed}_{H_2O}}$$

Temperature (°C)	Absolute Viscosity for H_2O (cp)
18	1.053
20	1.002
22	0.955
24	0.911
25	0.890
26	0.870
28	0.833

Conclusions

4. Inferring Conclusions
According to the invoice, the service station was supposed to receive equal amounts of SAE-10, SAE-20, SAE-30, SAE-40, SAE-50, and SAE-60 oil. Given that the oils with the lower SAE ratings have lower relative viscosities, infer which oil samples correspond to the SAE ratings indicated.

5. Organizing Information
Prepare a graph with flow time at room temperature on the y-axis and SAE rating on the x-axis.

6. Organizing Information
Prepare a graph with density on the y-axis and SAE rating on the x-axis.

7. Organizing Information
Prepare a graph with viscosity at room temperature on the y-axis and SAE rating on the x-axis.

8. Organizing Information
Prepare a graph with viscosity at room temperature on the y-axis and density on the x-axis.

9. Inferring Conclusions
How does temperature affect the viscosity of each sample?

10. Interpreting Graphics
Is there a relationship between density and viscosity?

11. Interpreting Graphics
What is the relationship between SAE rating and viscosity?

12. Interpreting Graphics
What is the relationship between viscosity and flow time?

Extensions

1. Predicting Outcomes
Estimate what flow times you would measure at each temperature if you repeated the tests in this lab with SAE-35 oil.

2. Relating Ideas
Malcolm is trying to get the last of the pancake syrup out of a bottle. What can he do to make the syrup come out of the bottle faster? Explain how your plan will take advantage of viscosity.

3. Research and Communication
Contact a manufacturer of lubrication products such as Valvoline or Pennzoil, and write a short paper on the development and properties of the oils used in this investigation.

Viscosity—New Lubricants

February 21, 1995

Ms. Sandra Fernandez
Director of Development
CheMystery Labs, Inc.
52 Fulton Street
Springfield, VA 22150

Dear Ms. Fernandez:

Our company manufactures a wide variety of industrial lubricants. We produce lubricating oils used in land vehicles, aircraft, and ships. Our range of products includes hydraulic fluids, engine oils, and lubricating grease.

We are presently interested in developing two lubricants for use in an experimental, high-rpm engine to be used in a pump. We want your firm to research and develop two multigrade oils, which are mixtures of other oils. One of these mixtures should have a relative viscosity of 20 cp at approximately 5°C. The second mixture should have a relative viscosity of 10 cp at approximately 21°C.

Once you test and identify the best method for mixing the oils to achieve the desired properties, we are prepared to manufacture and market these lubricants on a large scale. In exchange for your work, we are willing to pay up to $150 000.

Sincerely,

Judy Showalter

Judy Showalter
Director of Research
General Lubricants

Memorandum

Date: February 22, 1995
To: John X. Muhammad
From: Sandra Fernandez

CheMystery Labs, Inc.
52 Fulton Street
Springfield, VA 22150

We are still paying off several business loans and can certainly use this contract from General Lubricants. It will be your team's task to determine the ideal formulation of oils to achieve the two desired viscosities. Because of the number of research teams working on this project and the limited supply of oils available, you are limited to 50 mL of any single grade. Remember to work carefully to produce quality results.

Before you begin the job, I need to review the following items from you.
- detailed one-page plan for the procedure and all necessary data tables
- four proposed formulations for the new oils (Hint: you may wish to refer to data from Exploration 6-2 as you propose formulations.)
- detailed list of the equipment and materials you will need, along with the individual and total costs

Once your plan is approved, you may begin your lab work. Make at least 20 mL of each of the oil mixtures. When completed, prepare a report in the form of a two-page letter to Ms. Showalter that includes the following items.
- suggestion for multigrade oil to achieve the desired viscosities
- detailed and organized analysis of your test oil: density of samples, ratios of each grade used in samples, flow time, approximate SAE rating, and the average relative viscosity for each sample
- summary paragraph describing how you solved this problem
- detailed and organized data table
- detailed invoice including costs for equipment, supplies, and services

Required Precautions

 Always wear goggles and a lab apron to provide protection for your eyes and clothing.

 Do not touch any chemicals. If you get a chemical on your skin or clothing, wash it off at the sink while calling to your teacher.

 When you use a Bunsen burner, confine any long hair and loose clothing. Do not heat glassware that is broken, chipped, or cracked. Use tongs or a test-tube holder to handle heated glassware and other equipment. Because the oil tested in this lab is flammable, it should never be heated directly over a flame. Instead, use a hot-water bath, and never heat it above 60°C.

 Pins are sharp; use with care to avoid cutting yourself or others.

References

Refer to page 196 for information on intermolecular forces in nonpolar molecules, such as those found in oils. The viscosity testing procedure your team recently completed will be useful here. The data you gathered at that time can be used to help you select a promising mix of oils that should have properties close to those specified in the letter. You may need to calculate some relative viscosities for the pure oils using the Exploration's data for temperatures other than room temperature. The viscosity of water at 5°C is 1.52 cp, and at 21°C, it is 0.98 cp. Calculations of density are discussed on pages 55 and 56.

Spill Procedures/ Waste Disposal Methods

- In case of spills, follow your teacher's directions. Any oil spills should be wiped up with paper towels.
- Any unmixed oil sample should be placed in its own designated container.
- Put samples of mixed oil in the disposal containers designated by your teacher.
- Do not pour any oil samples down the drain or place them in the trash can.

MATERIALS for General Lubricants

REQUIRED ITEMS

(You must include all of these in your budget.)

Lab space/fume hood/utilities	15 000/day
Standard liquid disposal fee	20 000/L

REAGENTS AND ADDITIONAL EQUIPMENT

(Include in your budget only what you will need.)

Ice	500
Oil samples (SAE-10, -20, -30, -40, -50, and -60)	1 000/mL
50 mL beaker	1 000
400 mL beaker	1 000
25 mL graduated cylinder	1 000
Aluminum foil	1 000/cm²
Bunsen burner/related equipment	10 000
Duct tape	500/cm
Filter paper	500/piece
Glass stirring rod	1 000
Hot plate	8 000
Pin	500
Pipet	1 000
Ring stand/ring/wire gauze	2 000
Ruler	500
Six test tubes/holder/rack	2 000
Stopwatch	5 000
String	500/cm
Thermistor probe — LEAP	2 000
Thermometer clamp	1 000
Thermometer, nonmercury	2 000
Wax pencil	500
Weighing paper	500/piece

* No refunds on returned chemicals or unused equipment.

FINES

OSHA safety violation	2 000/incident

Chemical Reactions and Solid Fuel

Objectives

Recognize **several different reaction types.**

Write **balanced chemical equations for reactions observed in the laboratory.**

Prepare **a gel, and determine whether it is homogeneous or heterogeneous.**

Calculate **the mass of a fuel before burning and of the residue remaining after burning.**

Evaluate **the lab-produced solid fuel by measuring how well it heats water compared to a commercially available fuel.**

Situation

An entrepreneur, Ms. Toietta LaFaye, has contracted your company to research a new camping/emergency fuel. The idea came from an incident in which several campers forgot to pack the fuel for their camp stove. Because the area prohibited the use of campfires, the campers needed to use an alternative fuel source. One of the campers made a gel that they used as solid fuel. Some chalk was crushed and mixed with vinegar. This mixture was filtered through a napkin, and the filtrate was heated using a solar reflector. Some rubbing alcohol was poured into the solution to form a gel, which burned.

Ms. LaFaye wants to provide campers with a ready-made fuel. She wants you to prepare the fuel under controlled conditions in the laboratory to determine how and why it works, compare the product to a commercially available solid fuel, and identify what other products are produced as the fuel is made and used.

Background

One way to understand this process is to break it down into individual chemical reactions, as shown below. Ingredients are identified with placeholder names that can be replaced after they have been identified. Chalk is made mostly of calcium carbonate, $CaCO_3$. One of the main ingredients of vinegar is acetic acid, CH_3COOH. Rubbing alcohol is isopropanol, $CH_3CHOHCH_3$. It could be the flammable isopropanol in the gel that causes it to burn.

$$CaCO_3(s) + CH_3COOH(aq) \longrightarrow \text{Solution of Compound V} + \text{??}$$

$$\text{Compound V}(aq) + CH_3CHOHCH_3(l) \longrightarrow \text{Gel W} + \text{??}$$

$$\text{Gel W burns in air} \longrightarrow \text{Ash X}(s) + \text{??}$$

Ms. LaFaye also wants you to research whether ethanol, CH_3CH_2OH, would work better than isopropanol in the mixture.

Problem

In order to provide Ms. LaFaye with a complete report that gives her all of the information she needs, you must do the following.
- Prepare two kinds of gel in the lab: one with ethanol and one with isopropanol, and measure the mass of the gel produced.
- Compare this gel to commercially available fuels by measuring how hot they make similar amounts of water.
- Measure the mass of ash remaining after complete burning.
- Use what you know about chemical reactions from Chapter 7 to classify and describe each reaction.

Materials

- Chalk, 25 g
- Commercially available solid fuel
- Distilled water
- Ethanol, CH_3CH_2OH, 30 mL
- Isopropanol, $CH_3CHOHCH_3$, 30 mL
- Vinegar (acetic acid solution), 100 mL
- 250 mL beakers, 6
- 100 mL graduated cylinder
- Balance
- Crucible tongs
- Evaporating dishes, 2
- Glass stirring rod
- Mortar and pestle
- Ring stands with ring clamps, 3
- Wire gauze with ceramic center, 3

Gravity filtration option

- Glass funnel
- Filter paper

Vacuum filtration option

- Aspirator for spigot
- Büchner funnel
- Filter paper
- One-hole rubber stopper or sleeve
- Vacuum flask and tubing

Bunsen burner option

- Bunsen burner and related equipment

Hot plate option

- Hot plate

LEAP System option

- Thermistor probes, 3

Thermometer option

- Thermometers, non-mercury, 3

Safety

 Always wear goggles and a lab apron to provide protection for your eyes and clothing. If you get a chemical in your eyes, immediately flush it out at the eyewash station while calling to your teacher. Know the locations of the emergency lab shower and eyewash and how to use them.

 Do not touch any chemicals. If you get a chemical on your skin or clothing, wash it off at the sink while calling to your teacher. Make sure you carefully read the labels and follow the directions on all containers of chemicals that you use. Do not taste any chemicals or items used in the laboratory. Never return leftovers to their original containers; take only small amounts to avoid wasting supplies.

 When you use a Bunsen burner, confine any long hair and loose clothing. Do not heat glassware or other equipment that is broken, chipped, or cracked. Use tongs or a hot mitt to handle heated glassware and other equipment because hot glassware does not always look hot. If your clothing catches on fire, WALK to the emergency lab shower, and use it to put out the fire.

 Both ethanol and isopropanol are flammable. While working with ethanol and isopropanol to make the gelled or solid fuels, make sure there are no flames or other sources of ignition anywhere in the room. You may not obtain or work with ethanol or isopropanol until after all groups have finished heating the chalk-vinegar mixture and have turned off their burners or hot plates and allowed them to cool. No tests of the solid fuels should be done until all lab groups are finished making their fuels, all excess ethanol and isopropanol is stored in closed containers in the hood, the hood window is closed, and the hood fan is on.

 Always clean up the lab and all equipment after use, and dispose of substances according to proper disposal methods. Wash your hands thoroughly before you leave the lab after all lab work is finished.

Preparation

1. Organizing Data

Prepare two data tables in your lab notebook. The first data table should have three columns and four rows. Label the boxes in the first row *Measurement, Ethanol fuel,* and *Isopropanol fuel.* In the first column, label the second through fourth rows *Mass of empty evaporating dish (g), Mass of gel + dish (g),* and *Mass of ash + dish (g).*

2. *Organizing Data*

The second data table should have four columns and seventeen rows. Label the boxes in the first row *Time (min)*, *Temp. (°C)—ethanol fuel*, *Temp. (°C)—isopropanol fuel*, and *Temp. (°C)—commercial solid fuel*. Label the second through seventeenth boxes in the first column with the numbers 0–15. You will also need room below the data table to record your observations about the synthesis of the fuel and how it burned.

3. Measure the masses of the evaporating dishes to the **nearest 0.01 g.** Designate one to be used for the gelled isopropanol fuel and the other for the gelled ethanol fuel. Handle these dishes only with crucible tongs, even when cool, so that you will not cause errors by adding oil or dirt to them.

Technique

4. Place about 25 g of chalk in a mortar and pestle. Using a gentle circular motion, grind it into small pieces.

5. Completely transfer the chalk to a 250 mL beaker. Add about 100 mL of vinegar to the beaker, and stir for 5 to 8 min. Record your observations below your data table.

6. Set up your filtering apparatus. If you are using a Büchner funnel for vacuum filtration, follow step **6a.** If you are using a glass funnel for gravity filtration, follow step **6b.**

   a. *Vacuum filtration*

   Screw an aspirator nozzle into the faucet. Attach a piece of thick plastic tubing to the side arm of this nozzle. Attach the other end of the thick plastic tubing to the side arm of the filter flask. Place a one-hole rubber stopper on the stem of the funnel, and fit the stopper snugly in the neck of the filter flask, as shown in the illustration. Place the filter paper on the bottom of the funnel, so that it is flat and covers all of the holes.

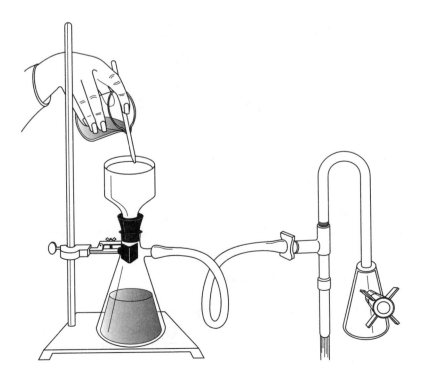

b. Gravity filtration

Use a glass funnel with filter paper as shown in the illustration, supported by a ring stand and ring. Prepare the filter paper by folding it as shown in the illustration. Fit the filter paper in the funnel, and wet it with a little water. Gently but firmly press the paper against the sides of the funnel so that no air can get between the funnel and the paper. The filter paper should not extend above the edge of the funnel.

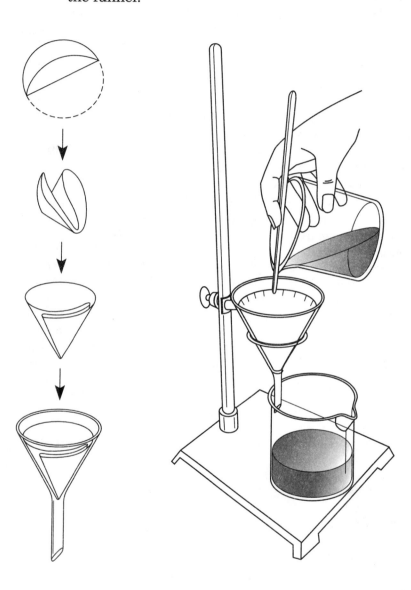

7. Filter the chalk-vinegar mixture. If you are using a Büchner funnel for vacuum filtration follow step **7a.** If you are using a glass funnel for gravity filtration, follow step **7b.**

a. Vacuum filtration

Prepare the filtering apparatus by pouring approximately 100 mL of water through the filter paper. After the water has gone through the funnel, empty the filter flask into the sink. Reconnect the filter flask, and pour the chalk-vinegar mixture into the funnel, using a stirring rod, as shown in the illustration.

b. Gravity filtration

Place a 250 mL beaker at the bottom of the glass funnel. Prepare the filtering apparatus by pouring approximately 100 mL of water through the filter paper. Pour the water that is in the beaker after passing through the filter into the sink. Place the beaker back at the bottom of the glass funnel so that it will collect the filtrate from the funnel. Pour the chalk-vinegar mixture into the funnel, using a stirring rod, as shown in the illustration.

8. Pour about 40 mL of the clear filtrate into the ethanol fuel evaporating dish. If you will be performing the lab during two days, reserve another 40 mL of the filtrate for later. If you will be performing the lab all in one day, pour another 40 mL of the clear filtrate into the isopropanol fuel evaporating dish. Remember to use crucible tongs to hold and carry the evaporating dishes. Do not use your hands.

9. Gently heat the contents of the evaporating dish or dishes for a few minutes with a hot plate or a Bunsen burner until the volume is reduced by half. Turn off the hot plate or burner when finished, and allow it to cool. **Do not proceed with the next step until your teacher tells you to continue.**

10. Add 30 mL of ethanol to the ethanol fuel evaporating dish. The gel should form in about 8 s. Decant any excess ethanol into the ethanol disposal container. Measure the mass of the dish and gel. Record this and any observations about the gel and its synthesis in your data table. If you will be performing the lab during two days, proceed with step **13** today to test the ethanol-based fuel. Then, continue with step **11** when you are ready to make the isopropanol fuel.

11. Repeat step **10** using isopropanol to make the gelled fuel in the isopropanol fuel evaporating dish. Decant any excess isopropanol into the isopropanol disposal container. Measure the mass of the dish and gel. Record this and any observations about the gel and its synthesis in your data table.

12. **Do not proceed with the next step until all students have finished making their fuel, the disposal containers have been placed in the fume hood, and your teacher tells you to continue.**

13. Set up a ring stand and ring clamp for each fuel sample to be tested. Adjust each clamp so that its height is about 10.0 cm from the base of the ring stand. Place a wire gauze with ceramic center on each ring clamp. Set a 250 mL beaker on top of each gauze. Under one beaker, place the ethanol fuel. Under another, place the isopropanol fuel. Under the third, place the commercially available solid fuel.

14. Using a graduated cylinder, measure 75.0 mL of distilled water into each of the beakers above the fuel samples. (The volume does not have to be exactly 75.0 mL, but the volume of water in each beaker must be the same.)

15. Using three thermometer clamps and three thermometers or LEAP System thermistor probes, suspend a thermometer or thermistor probe in each beaker so that it does not touch the sides or bottom of the beaker. Record the initial temperature of all of the water samples. If they are not all the same, wait a few minutes until they are.

16. When you are ready to continue, ask your teacher to light the fuels for you. Record the temperature of the water in each beaker at 1.0 min intervals for 15 min or until the fuels burnout.

17. Wait until the fuels have stopped burning and the evaporating dishes have cooled thoroughly, and then measure the mass of the ash and record it in your data table.

Cleanup and Disposal

18. Place the ash in its own disposal container. Any excess alcohol should already be in the ethanol disposal container or the isopropanol disposal container. The filter paper and the solid material on it may be placed in the trash, and any excess filtrate from the vinegar-chalk mixture may be poured down the drain. Always wash your hands thoroughly after cleaning up the area and equipment.

Analysis and Interpretation

1. **Analyzing Observations**
   Explain what observations indicated that a chemical reaction was taking place when you added the vinegar to the chalk.

2. **Analyzing Information**
   Look at the first equation in the Background section at the beginning of this Exploration. Write a balanced equation for this reaction, including formulas for the products. (Hint: the reaction is a double-displacement reaction. When acetic acid is dissolved in water, some of it dissociates into acetate ions and hydrogen ions.)

3. **Analyzing Information**
   One of the products from the equation you wrote in item **2** underwent a decomposition reaction. One of the products was water. Write the balanced chemical equation for this decomposition reaction. How does this equation relate to your observations in item **1**?

4. **Analyzing Observations**
   Was there any evidence that the heating of the filtrate caused a chemical reaction?

5. **Analyzing Observations**
   Was there any evidence that the addition of the alcohols to form a gel (the second equation listed in the Background section) caused a chemical reaction?

6. **Analyzing Observations**
   Is the gel a homogeneous or heterogeneous mixture? Explain your answer.

7. **Analyzing Information**
   Assuming that the part of the gels that actually burned was the alcohol, write balanced equations for each combustion reaction.

8. **Analyzing Observations**
   Is the ash that remains after burning your gels represented as a product in the chemical equations written in item **7**? What is the primary component of the ash?

9. Organizing Data

How many grams of the component described in item **8** were contained in each gel sample? Calculate the corresponding percentages by mass.

10. Organizing Data

How many grams of each alcohol were present in the gel samples? How many moles of each alcohol were present?

11. Predicting Outcomes

The amount of energy produced by the combustion of exactly one mole of ethanol has been measured to be about − 1367 kJ. The combustion of one mole of isopropanol releases about − 2006 kJ. Calculate how much energy should have been produced as the gelled fuels burned. According to these figures, should more energy have been provided by your isopropanol-based fuel or your ethanol-based fuel?

12. Organizing Data

Graph the results of the heating tests for all three fuels on the same graph, with time in minutes on the x-axis and temperature in °C on the y-axis.

Conclusions

13. Interpreting Graphics

Using your graph from item **12,** which fuel was most effective at heating water? Which was least effective?

14. Evaluating Conclusions

Can you think of reasons why using the fuel that burns the fastest and hottest might not be the best choice for this kind of product? (Hint: consider that the fuel is designed to be safely and conveniently carried by campers and hikers.)

15. Inferring Conclusions

Ms. LaFaye is concerned about pollution issues related to both the manufacture and the use of the gelled fuel. In particular, she is concerned about the production of pollutants such as nitrogen oxide compounds and hydrocarbons. Are these likely to be produced when the fuel is manufactured or burned? Explain your answers.

Extensions

1. Applying Conclusions

Calculate the amount of temperature change in 75 mL of H_2O per gram of gelled fuel. What mass of each gelled or solid fuel would be necessary to heat a cup (250 mL) of H_2O from 21°C to 100°C?

2. Evaluating Methods

If Ms. LaFaye makes fuel chunks just big enough to heat a cup of H_2O to 100°C, as described in Extensions item **1,** what is the unit cost for a chunk of each type of fuel?

Reagent	Cost
Chalk	$6.25/kg
Ethanol	$3.27/L
Isopropanol	$4.62/L
Vinegar	$0.86/L

Specific Heat Capacity

Situation

Today's Housewares, Inc. is planning to introduce a new line of cookware. In deciding which metals to use for the cookware, they have to consider how well the metals retain heat. The metals being considered by the company are aluminum, iron, and copper. Your company has a contract with Today's Housewares, Inc. to test the specific heat capacity of each type of metal. This information, along with other considerations, will be used by Today's Housewares, Inc. to decide which metal to use.

Background

Heat is a form of energy that cannot be measured directly. However, changes in heat can be determined by measuring changes in temperature. To calculate the amount of heat absorbed or lost by a substance as it changes temperature, use the following equation, which is explained on pages 315–318 of the textbook.

$$\text{change in heat} = m \times c_p \times \Delta t$$

Metals with high specific heats tend to heat up and cool down slowly because it takes more energy to raise their temperature.

Measurements in a calorimeter are based on the assumption that the heat absorbed by the water in the calorimeter is the same as the heat released as the metal cools down. However, some heat is lost because it is impossible to insulate perfectly. The problem of heat loss is usually solved by calibrating the calorimeter to determine what adjustments are necessary to compensate for the heat energy that is lost.

Problem

To get the information for your report to Today's Housewares, Inc., you will need to do the following.
- Make and calibrate a calorimeter using carefully measured masses and temperatures of warm and cold water.
- Measure the mass of each metal sample and heat each one, measuring its temperature.
- Add the heated metal to a carefully measured mass and temperature of cold water in the calorimeter.
- Measure the final temperature of the water and metal to calculate the change in heat energy for the water.
- Determine the specific heat capacity of the metal after accounting for the adjustment specified by the calibration step.

Objectives

Build and calibrate a simple calorimeter.

Relate measurements of temperature to changes in heat energy.

Demonstrate proficiency in using calorimetry techniques to determine the specific heat capacities of metals.

Materials

- Aluminum metal sample
- Copper metal sample
- Iron metal sample
- 400 mL beakers, 2
- 100 mL graduated cylinder
- Balance
- Beaker tongs
- Boiling chips
- Glass stirring rod
- Plastic foam cups for calorimeter, 2
- Scissors
- Ring stand
- Test-tube holder
- Test tube, large or medium, 3

Bunsen burner option

- Bunsen burner and related equipment
- Ring clamp
- Wire gauze with ceramic center

Hot plate option

- Hot plate

LEAP System option

- LEAP System
- LEAP thermistor probe

Thermometer option

- Thermometer, non-mercury

Safety

 Always wear goggles and a lab apron to provide protection for your eyes and clothing. If you get a chemical in your eyes, immediately flush it out at the eyewash station while calling to your teacher. Know the locations of the emergency lab shower and eyewash station and how to use them.

 When you use a Bunsen burner, confine any long hair and loose clothing. Do not heat glassware that is broken, chipped, or cracked. Use tongs or a hot mitt to handle heated glassware and other equipment because hot glassware does not look hot. If your clothing catches on fire, WALK to the emergency lab shower, and use it to put out the fire.

 Scissors are sharp; use with care to avoid cutting yourself or others.

 Always clean up the lab and all equipment after use, and dispose of substances according to proper disposal methods. Wash your hands thoroughly before you leave the lab after all lab work is finished.

Preparation

1. **Organizing Data**
   Prepare a data table with four columns and four rows for the calibration of your calorimeter. Label the boxes in the top row *Measurement, Cool H₂O, Hot H₂O,* and *Calorimeter.* Label the second through fourth boxes in the first column *Volume (mL), Init. temp. (°C),* and *Final temp. (°C).*

2. **Organizing Data**
   Prepare another data table with four columns and seven rows for the testing of the metals. Label the boxes in the top row *Measurement, Al, Cu,* and *Fe.* Label the second through seventh boxes in the first column *Test-tube mass (g), Metal + test-tube mass (g), Metal init. temp. (°C), H₂O volume (mL), H₂O init. temp (°C),* and *Final temp. (°C).*

3. Construct a calorimeter from two plastic foam cups. Trim the lip of one cup, and use that cup for the top of the calorimeter. Use a pencil to make a small hole in the center of the base of this cup (the top of the calorimeter) so that a thermometer or thermistor can be inserted, as shown in the illustration on the next page. Place the calorimeter in one of the beakers to help prevent it from tipping over. If a thermistor is to be used, fasten it to the cup with a rubber band, as shown on page 720 of the textbook.

4. Label three test tubes *Al, Cu,* and *Fe.*

Technique

Calibrating the calorimeter

5. Use a graduated cylinder to measure about 50 mL of cold water. Record the exact volume to the nearest 0.1 mL in the calibration data table. (Note: place an "X" in the box on the *Volume (mL)* row in the column labeled *Calorimeter.* The volume of the calorimeter will not be used in this Exploration.)

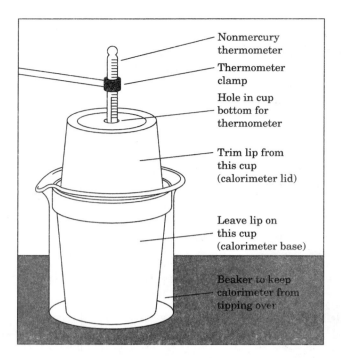

Nonmercury thermometer

Thermometer clamp

Hole in cup bottom for thermometer

Trim lip from this cup (calorimeter lid)

Leave lip on this cup (calorimeter base)

Beaker to keep calorimeter from tipping over

6. Pour the cold water into your calorimeter, and allow the water to reach room temperature. Measure the temperature every minute until the temperature stays the same for 3 min. Record this temperature as the initial temperature for both the cool water and the calorimeter.

7. If you are using a Bunsen burner, clamp a ring to the ring stand, and place the other beaker on the wire gauze on top of the ring clamp. If you are using a hot plate, place the beaker directly on the hot plate. Heat about 75 mL of water in the beaker to 70–80°C.

8. Using beaker tongs, remove the beaker from the heat source, and, using tongs or a hot mitt, pour about 50 mL of hot water into a graduated cylinder. Measure and record the volume to the nearest 0.1 mL in the calibration data table.

9. Using a glass stirring rod, stir the hot water in the graduated cylinder for about a minute and measure the temperature. Record this value in the calibration data table as the initial hot-water temperature. ***Never use a thermometer to stir anything. It will break easily. The glass wall surrounding the bulb is very thin to provide quick and accurate temperature readings.***

10. Immediately pour all of the hot water into the calorimeter with the cool water. Put on the lid and gently swirl the beaker in which the calorimeter rests for 30 s. Record the highest temperature attained by the mixture as the final temperature for the hot water, the cool water, and the calorimeter.

11. Empty the calorimeter. Pour the water down the drain.

Testing metal samples

12. Measure the masses of the empty test tubes, and record them in the second data table. Add the appropriate metal to each of the test tubes. The height of the metal in the test tube must be less than the height of the water in the calorimeter cup. Measure and record the mass of each test tube with metal.

13. Add about 300 mL of water and a few boiling chips to the beaker from step **7.** Place all three test tubes with metal in the beaker.

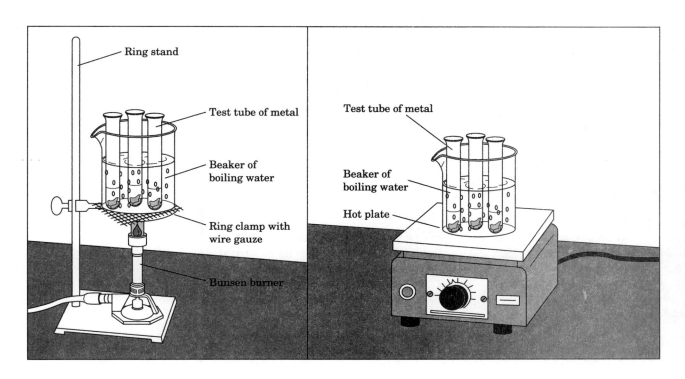

Ring stand

Test tube of metal

Beaker of boiling water

Ring clamp with wire gauze

Bunsen burner

Test tube of metal

Beaker of boiling water

Hot plate

14. Heat the beaker of water with the test tubes and metal using a hot plate or a Bunsen burner until the water begins to boil, and continue heating for approximately 10 min. Be sure that the metal remains below the surface of the water throughout the heating cycle. Add more water only if necessary.

15. While the water boils, measure about 75 mL of cold water in a graduated cylinder. Record the exact volume to the nearest 0.1 mL in the *Al* column in the second data table. Allow the water to come to room temperature, as in step **6.** Record this value as the initial water temperature in the second data table.

16. Measure and record the temperature of the boiling water after 10 min. You can assume that each metal is at the same temperature as the boiling water. Record the value in the second data table as the initial temperature for each metal. Keep the water boiling through the following steps.

17. First make certain that the thermometer is not in the calorimeter because adding the metal could cause it to break. Then remove the *Al* test tube from the hot-water bath with a test-tube holder because it is hot enough to cause burns. Transfer the aluminum metal to the calorimeter. Quickly and carefully put the top on the calorimeter and gently insert the thermometer into the water. Gently swirl the beaker containing the calorimeter for 30 s, making sure that the metal does not hit the thermometer. Monitor the temperature, and record in your data table the highest temperature attained by the water.

18. Remove the aluminum sample from the calorimeter, and pour the water down the drain. Place the aluminum sample on a paper towel. When it is dry, put it in the appropriate disposal container.

19. Repeat steps **15** and **17–18** for samples of copper and iron.

Cleanup and Disposal

20. Turn off the boiling water and allow it to cool. Place the dried metals in separate disposal containers designated by your teacher. Clean up your equipment and lab station. After the water has cooled, pour it down the drain. Always wash your hands thoroughly before you leave the lab after cleaning up the lab area and equipment.

Analysis and Interpretation

1. Organizing Ideas
State the scientific law that is the basis for the assumption that the heat energy lost by the metal as it cools will be equal to the heat energy gained by the water and the calorimeter.

2. Organizing Data
Assuming that 1.00 g/mL is the density of cold water and 0.97 g/mL is the density of hot water, calculate the masses of cold and hot water used in the calibration step.

3. Analyzing Information
Calculate Δt for the hot water and the cold water.

4. Applying Ideas
Using the value of the specific heat capacity of water, which is 4.180 J/g•°C, and the specific heat capacity equation given in the Background section, calculate the amounts of heat released by the hot water and absorbed by the cold water.

5. Evaluating Data
Explain why the two values you calculated in item **4** should be slightly different. How much heat energy was released by the calorimeter to the surroundings?

6. Applying Ideas
To be able to use a calorimeter for many different temperature changes without calibrating every time, chemists often determine the heat capacity of the calorimeter, or C', from their calibration data, according to the following equation.

$$C' = \frac{\text{heat lost to surroundings}}{\Delta t_{calorimeter}}$$

Determine C' for your plastic foam cup calorimeter.

7. Evaluating Methods
Explain whether it is best to have a high or a low value for the heat capacity of a calorimeter.

8. Organizing Data
Calculate the mass of each metal used.

9. Organizing Data
Calculate the temperature changes, Δt, for the water and for the metal in each of the three tests.

Conclusions

10. Organizing Ideas
Write a valid equation involving these three quantities: heat energy released by the metal as it cools, heat energy absorbed by the water as it warms up, and heat energy released to the surroundings by the calorimeter. (Hint: refer to item **1**.)

11. Applying Ideas
Identify which two of the quantities in the equation from item **10** can be calculated from values you already know, have measured, or have calculated in the previous items. (Hint: apply the specific heat capacity equation and the equation in item **6**.)

12. Applying Ideas
Set up three different calculations, one for each of the metals tested, to determine the values of the quantities identified in item **11**. Assume that the density of water is 1.00 g/mL.

13. Inferring Conclusions
Using the specific heat capacity equation and the answers from the previous items, determine the value of the specific heat capacity for each metal.

14. Evaluating Conclusions
Compare your values for specific heats to the values found in a handbook, such as the *CRC Handbook of Chemistry and Physics* or *Lange's Handbook of Chemistry*. (You may need to convert the values given from calories to joules. The conversion factor is: 1 cal = 4.184 J.) Calculate your percent error.

15. Organizing Ideas
Refer to your data to determine how the specific heat capacity of the metals compares to the specific heat capacity of water. Which would absorb the same amount of heat energy with a greater change in temperature—metal or water?

16. Evaluating Conclusions
Today's Housewares is trying to make pots and pans that will quickly reach cooking temperature. Explain, using your data, which metal is the best choice if this is the primary concern.

Extensions

1. Designing Experiments
Share your data with other lab groups, and calculate a class average for the different specific heat capacities of the metals. Compare the averages to the figures in a chemical handbook, and calculate the percent error for the class averages.

2. Applying Ideas
Design a different calorimeter with better insulation. Describe what tests you would use to calibrate it and measure its heat capacity, C'. If your teacher approves the design of the calorimeter and your plans for testing it, build the calorimeter and determine how much better it works. Compare this value to the value calculated in the Exploration, and describe the improvement you achieved as a percentage.

3. Applying Ideas
Today's Housewares, Inc. is trying to design the handles for its cookware. Should they select materials with high or low specific heat capacities? Explain your answer.

Specific Heat Capacity— Terrorist Investigation

TOP SECRET

March 13, 1995

Marissa Bellinghausen
Director of Investigations
CheMystery Labs, Inc.
52 Fulton Street
Springfield, VA 22150

Dear Ms. Bellinghausen:

Yesterday morning, at 0600, CrestAir flight 998 crashed approximately 5 miles from Seymour Johnson AFB, near Goldsboro, NC. Approximately 15 minutes after takeoff from Raleigh-Durham Airport, eyewitnesses heard several loud explosions and saw the plane on fire and hurtling toward the ground. All 50 passengers and 8 crew members were killed on impact.

The Federal Aviation Administration's inspectors have recovered parts of what they believe was an incendiary device placed in the luggage compartment. Reliable sources inform us that several terrorist groups have threatened a wave of weekly airliner bombings. We must identify the responsible party quickly to avert any future calamities.

There are four possible terrorist groups that could have been responsible for this bombing. Each group is known for a specific metal from which they construct their devices.

Popular Front for the Liberation of Fredonia: lead
Penguins' Patrol: zinc
Spleen Activist Group: aluminum
Drachmanian Separatists: iron

Inform us as soon as you can supply evidence to implicate a group.

Sincerely,

Maxwell Stone

Maxwell Stone
Director of Anti-Terrorist Operations
North American Theater
Central Intelligence Agency

Memorandum

Date: March 14, 1995
To: Thi Antoan
From: Marissa Bellinghausen

CheMystery Labs, Inc.
52 Fulton Street
Springfield, VA 22150

For this project, you will be provided with a sample of the <u>metal from the crash site</u>, but it is <u>not very large</u>. Because the CIA hopes to eventually arrest the guilty party, the sample you use <u>must remain intact</u> so that it can be <u>evidence for a later trial</u>. The CIA wants an <u>ironclad case</u>, so check at least one other physical property besides the <u>specific heat capacity</u> that will distinguish these metals. You may need to <u>check reference works</u> for information on the metals that could have been used. Remember, we have <u>less than 72 hours</u> to identify the terrorist group responsible.

The CIA will reimburse us for <u>reasonable expenses</u>, but I want to keep track of how you plan to spend the money, so I <u>need the following items</u> from you <u>before you begin work</u>.

- detailed one-page <u>plan</u> for the procedure, indicating what <u>properties you will measure</u> and how, and all necessary <u>data tables</u>
- detailed list of all <u>equipment and materials</u> you will need, along with individual and total <u>costs</u>

After you complete your work, prepare a two-page report in the form of a <u>letter to Mr. Stone</u>. You must include the following.

- name of the <u>terrorist organization implicated</u>
- summary paragraph describing <u>how you solved this problem</u>
- detailed and organized <u>data and analysis section</u>, including calculations for <u>specific heat</u> of the sample, some <u>other physical property</u>, and averages of <u>multiple trials</u>
- <u>detailed invoice</u> for services and expenses, charged to "Operation Bird-in-the-Hand" (Remember that the invoice will have to get past the CIA's <u>new auditor-general</u>, who is rumored to be a <u>stickler for details</u>.)

Required Precautions

Always wear goggles and a lab apron to provide protection for your eyes and clothing.

When you use a Bunsen burner, tie back loose clothing and hair. Do not heat glassware that is broken, chipped, or cracked. Use tongs whenever handling glassware or other equipment that has been heated.

Do not touch any chemicals. If you get a chemical on your skin or clothing, wash it off at the sink while calling to your teacher.

Always wash your hands thoroughly when finished.

Spill Procedures/ Waste Disposal Methods

- Use paper towels to dry the metal samples when you have finished.
- Place the dried metal samples in the containers designated by your teacher.

References

Refer to pages 315–316 for more information on the specific heat capacities of metals. There are many ways to identify metals based on their physical properties, such as measuring magnetic effects, conductivity, or density. The procedure for measuring specific heat capacity is very similar to a procedure your team recently completed comparing the specific heat capacities of several metals. Your data from that procedure could be useful in helping you identify the metal from the bomb.

MATERIALS for CIA Operation Bird-in-the-Hand

REQUIRED ITEMS
(You must include all of these in your budget.)

Lab space/fume hood/utilities	15 000/day
Standard disposal fee	2 000
Balance	5 000
Beaker tongs	1 000
Chemical handbook with metal properties	500

REAGENTS AND ADDITIONAL EQUIPMENT
(Include in your budget only what you'll need.)

100 mL graduated cylinder	1 000
400 mL beaker	2 000
Boiling chips	500
Bunsen burner	10 000
Desiccator	3 000
Filter paper	500/piece
Glass stirring rod	1 000
Hot plate	8 000
Magnets	2 000
Plastic foam cups	1 000
Ring stand/ring/wire gauze	2 000
Scissors	500
Spatula	500
Stopwatch	5 000
Test tube, large	1 000
Test-tube holder	500
Thermistor probe — LEAP	2 000
Thermometer, nonmercury	2 000
Wash bottle	500
Weighing paper	500/piece

* No refunds on returned chemicals or unused equipment.

FINES

OSHA safety violation	2 000/incident

9-2

Technique Builder

Constructing a Heating/ Cooling Curve

Objectives

Observe the temperature and phase changes of a pure substance.

Measure the time needed for the melting and freezing of a specified amount of substance.

Graph experimental data and determine the melting and freezing points of a pure substance.

Analyze the graph for the relationship between melting point and freezing point.

Identify the relationship between temperature and phase change for a substance.

Infer the relationship between heat energy and phase changes.

Recognize the effect of an impurity on the melting point of a substance.

Analyze the relationship between energy, entropy, and temperature.

Situation

Sodium thiosulfate pentahydrate, $Na_2S_2O_3 \cdot 5H_2O$, is produced by a local manufacturing firm and sold nationwide to photography shops, paper processing plants, and textile manufacturers. Purity is one condition of customer satisfaction, so samples of $Na_2S_2O_3 \cdot 5H_2O$ are taken periodically from the production line and tested for purity by an outside testing facility. Your company has been tentatively chosen because your proposal was the only one based on melting and freezing points rather than the more expensive titrations with iodine. To make the contract final, you must convince the manufacturing firm that you can establish accurate standards for comparison.

Background

As heat energy flows from a liquid, its temperature drops. The entropy, or random ordering of its particles, also decreases until a specific ordering of the particles results in a phase change to a solid. If heat energy is being released or absorbed by a substance remaining at the same temperature, this is evidence that a dramatic change in entropy, such as a phase change, is occurring. Because all of the particles of a pure substance are identical, they all freeze at the same temperature, and the temperature will not change until the phase change is complete. If a substance is impure, the impurities will not lose thermal energy in the same way that the rest of the particles do. Therefore, the freezing point will be somewhat lower, and a range of temperatures instead of a single temperature.

Problem

To evaluate the samples, you will need a heating/cooling curve for pure $Na_2S_2O_3 \cdot 5H_2O$ that you can use as a standard. To create and use this curve, you must do the following.

• Obtain a measured amount of pure $Na_2S_2O_3 \cdot 5H_2O$.

• Melt and freeze the sample, periodically recording the time and temperature.

• Graph the data to determine the melting and freezing points of pure $Na_2S_2O_3 \cdot 5H_2O$.

• Interpret the changes in energy and entropy involved in these phase changes.

• Verify the observed melting point against the accepted melting point found in reference data from two different sources.

• Use the graph to qualitatively determine whether there are impurities in a sample of $Na_2S_2O_3 \cdot 5H_2O$.

Safety

Always wear goggles and a lab apron to provide protection for your eyes and clothing. If you get a chemical in your eyes, immediately flush it out at the eyewash station while calling to your teacher. Know the locations of the emergency lab shower and eyewash station and how to use them.

Do not touch any chemicals. If you get a chemical on your skin or clothing, wash it off at the sink while calling to your teacher. Make sure you carefully read the labels and follow the directions on all containers of chemicals that you use. Never return leftovers to their original containers; take only small amounts to avoid wasting supplies.

When you use a Bunsen burner, confine any long hair and loose clothing. Do not heat glassware that is broken, chipped, or cracked. Use tongs or a hot mitt to handle glassware or other equipment that has been heated because hot equipment does not always look hot. If your clothing catches on fire, WALK to the emergency lab shower and use it to put out the fire.

Always clean up the lab and all equipment after use, and dispose of substances according to proper disposal methods. Wash your hands thoroughly before you leave the lab after all lab work is finished.

Preparation

1. **Organizing Data**
   Prepare spaces in your lab notebook to record the elapsed time and approximate temperature. Prepare a data table in your lab notebook with 6 columns and at least 25 rows. Label the first three boxes in the top row *Time (s)*, *Temp. (°C)*, and *Observations of cooling*. Label the fourth through sixth boxes in the top row *Time (s)*, *Temp. (°C)*, and *Observations of heating*.

2. Fill two 600 mL beakers three-fourths full of tap water.

3. Heat water for a hot-water bath. If you are using a Bunsen burner, attach to a ring stand a ring clamp large enough to hold a 600 mL beaker. Adjust the height of the ring until it is 10 cm above the burner. Cover the ring with wire gauze. Set one 600 mL beaker of water on the gauze. If you are using a hot plate, rest the beaker of water directly on the hot plate.

4. Monitor the temperature of the water with a thermometer or LEAP thermistor probe. Complete steps **5–8** while the water is heating.

5. Cool the water for a cold-water bath. Fill a small plastic washtub with ice. Form a hole in the ice that is large enough for the second 600 mL beaker. Insert the beaker and pack the ice around it up to the level of the water in the beaker.

Materials

- Ice
- $Na_2S_2O_3 \cdot 5H_2O$
- 600 mL beakers, 3
- Balance, centigram
- Beaker tongs
- Chemical reference books
- Forceps
- Graph paper
- Hot mitt
- Plastic washtub
- Ring stands, 2
- Ring clamps, 3
- Ruler
- Stopwatch or clock with a second hand
- Test-tube clamp
- Test tube, Pyrex, medium
- Thermometer clamp
- Wire gauze with ceramic center, 2
- Wire stirrer

Bunsen burner option
- Bunsen burner
- Gas tubing
- Striker

Hot plate option
- Hot plate

LEAP System option
- LEAP System
- Thermistor probes, LEAP, 2

Thermometer option
- Thermometers, non-mercury, 2

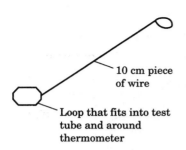

10 cm piece of wire

Loop that fits into test tube and around thermometer

6. Bend the piece of wire into the shape of a stirrer, as shown in the illustration. One loop should be narrow enough to fit into the test tube, yet wide enough to easily fit around the thermometer without touching it.

7. Prepare the sample. Assemble the test tube, thermometer, and stirrer, as shown in the illustration. Attach the entire assembly to a second ring stand. Then, add enough $Na_2S_2O_3 \cdot 5H_2O$ crystals so that the test tube is about one-quarter full and the thermometer bulb is well under the surface of the crystals.

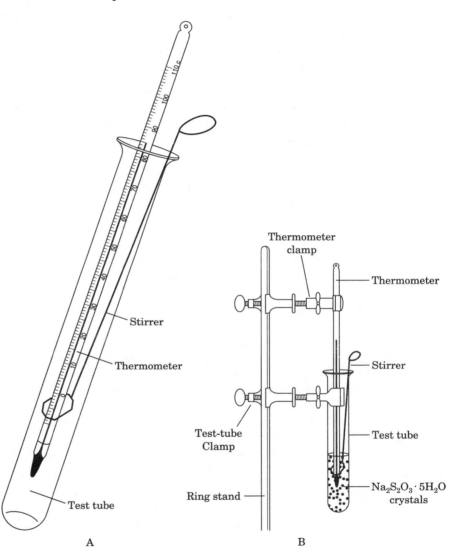

Stirrer

Thermometer

Test tube

A

Thermometer clamp

Thermometer

Stirrer

Test tube

Test-tube Clamp

Ring stand

$Na_2S_2O_3 \cdot 5H_2O$ crystals

B

8. Set up the container for the hot-water bath. Attach two ring clamps, one above the other, to the second ring stand beneath the test-tube assembly. Place a wire gauze with ceramic center on the lower ring. Set a third 600 mL beaker, which should be empty, on the gauze and raise the beaker toward the test-tube assembly until it surrounds nearly one-half of the tube's length. The beaker will pass through the ring clamp without gauze, and the test tube should not touch the bottom or sides of the beaker, as shown in the illustration on the next page. The top clamp keeps the beaker from tipping when the beaker is filled with the hot water.

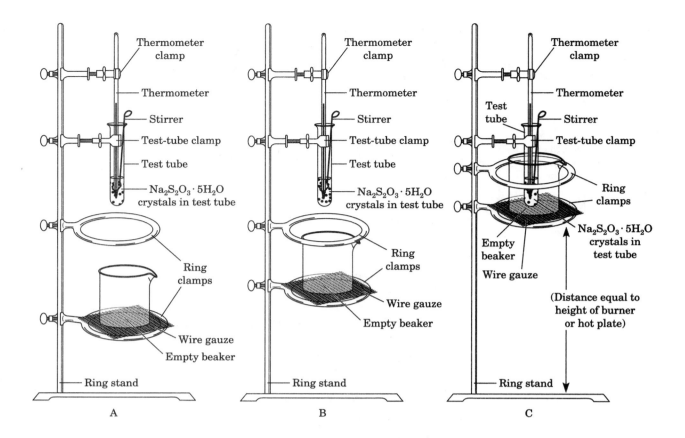

A   B   C

Technique

Melting a Solid—Quick Test

9. Check the temperature of the water for the hot-water bath. When it is 85°C, turn off the burner or hot plate. If the temperature is already greater than 85°C, shut off the burner or hot plate, and add a few pieces of ice to bring the temperature down to 85°C. Then, using beaker tongs, remove the beaker of hot water from the burner. Using tongs or a hot mitt carefully pour the water into the empty beaker until the water level is well above the level of the solid inside the test tube. Set the empty beaker on the counter. You will use it again in step **20.**

10. Begin timing. The second the water is poured, one member of the lab group should begin timing, while the other reads the initial temperatures of the bath with one thermometer or LEAP thermistor and sample with the other thermometer or LEAP thermistor.

11. Occasionally stir the melting solid by gently moving the stirrer up and down. Be careful not to break the thermometer bulb. Monitor the temperature of the $Na_2S_2O_3 \cdot 5H_2O$ and the hot-water bath with separate thermometers or probes.

12. When the temperature of the liquid $Na_2S_2O_3 \cdot 5H_2O$ is approximately the same as that of the hot-water bath, stop timing. Record the final temperature of the liquid $Na_2S_2O_3 \cdot 5H_2O$ and the elapsed time in your lab notebook. This temperature is the approximate melting point of your sample. Knowing this value can help you make the careful observations necessary to determine a more precise value.

13. Using a hot mitt, hold the beaker of hot water with one hand while using the other hand to gently loosen only the lower ring clamp enough so that the beaker of hot water can be lowered and removed. Remove the beaker of hot water, set it on the gauze above the burner, and let it reheat to 65°C while you perform steps **14–20.**

Freezing a Liquid

14. Set up the cold-water bath. Remove the beaker of cold water from the ice and place it on the ring with the gauze, well below the test tube. Steady the beaker with one hand while raising it until the level of the cold water is well above the level of the liquid inside the test tube. The test tube should not touch the bottom or sides of the beaker.

15. Begin timing. The second that the cold water is in place, one member of the lab group should begin timing, while the other reads the initial temperatures of the sample and the bath. Record the initial time and temperatures in the left half of the data table. The starting temperature of the liquid should be near 80°C.

16. Monitor the cooling process. Measure and record the time and the temperature of the $Na_2S_2O_3 \cdot 5H_2O$ every 15 s in the left half of the data table. Also record observations about the substance's appearance and other properties in the *Observations of cooling* column in the data table. When the temperature reaches 50°C, use forceps to add one or two seed crystals of $Na_2S_2O_3 \cdot 5H_2O$ to the test tube.

17. Continue taking temperature readings every 15 s, stirring continuously, until a constant temperature is attained. (A temperature is constant if it is recorded at four consecutive 15 s intervals.) **Do not try to move the thermometer, thermistor probe, or stirrer when solidification occurs.**

18. Finish timing. Continue taking readings until the temperature of the solid differs from the temperature of the cold-water bath by 5°C.

19. Remove the cold-water bath. Grasp the beaker with one hand, carefully loosen its supporting ring clamp with the other hand, and lower the beaker of cold water away from the test tube. Remove the beaker from the ring and set it on the counter.

Melting a Solid

20. Set up the container for the hot-water bath. Place the empty beaker from step **9** on the ring and wire gauze. Steady the beaker as you raise it to surround the test tube as you did in step **8,** but this time allow room for the Bunsen burner to be placed under the beaker.

21. Fill the hot water bath. Use the second thermometer or LEAP thermistor to check the temperature of the water for the hot-water bath. When it is 65°C, turn off the burner or hot plate. If the temperature is greater than 65°C, add a few pieces of ice to lower the temperature. Using tongs or a hot mitt, carefully pour the hot water into the empty beaker until the water level is well above the level of the solid inside the test tube. Set the empty beaker on the counter.

22. Begin timing. The second that the water is poured, one member of the lab group should begin timing while another reads initial temperatures of the water bath and the solid $Na_2S_2O_3 \cdot 5H_2O$. Record the solid's temperature in the right half of the data table. The starting temperature of the solid should be below 35°C.

23. Maintain the bath's temperature. Move the burner or hot plate under the hot-water bath and continue heating the water in the bath. Adjust the position and size of the flame or the setting of the hot plate so that the temperature of the hot-water bath remains between 60°C and 65°C.

24. Monitor the warming process. Record the temperature of the sample every 15 s. Use the stirrer, when it becomes free of the solid, to gently stir the contents of the test tube. Also record observations about the substance's appearance and other properties in the *Observations of heating* column in the data table.

25. Continue taking readings until the temperature of the $Na_2S_2O_3 \cdot 5H_2O$ differs from that of the hot-water bath by 5°C.

26. Record the final temperature and the time.

27. Turn off the burner or hot plate.

Cleanup and Disposal

28. Remove the thermometer or thermistor probe from the liquid $Na_2S_2O_3 \cdot 5H_2O$ and rinse it. Pour the $Na_2S_2O_3 \cdot 5H_2O$ from the test tube into the disposal container designated by your teacher. If you used a Bunsen burner, check to see that the gas valve is completely turned off. Remember to wash your hands thoroughly after cleaning up the lab area and all equipment.

Analysis and Interpretation

1. **Organizing Data**
   Plot both the heating and cooling data on the same graph. Place the time (marked in 15 s intervals) on the x-axis and the temperature in degrees Celsius on the y-axis.

2. **Interpreting Graphics**
   Describe and compare the shape of the cooling curve with the shape of the heating curve.

3. **Interpreting Graphics**
   Locate the freezing and melting temperatures on your graph. Compare them and comment on why they have different names.

4. **Evaluating Methods**
   One purpose of the quick test for melting point is summarized in step 12. State this purpose and explain how it prepares you for steps 17 and 24.

5. **Evaluating Data**
   Compare your melting point with that found in references. What is your percent error?

HRW material copyrighted under notice appearing earlier in this work.

Conclusions

6. Analyzing Information
As the liquid cools, what is happening to the kinetic energy and the entropy of the following?
a. $Na_2S_2O_3 \cdot 5H_2O$
b. the water bath

7. Analyzing Information
What happened to the temperature of the sample from the time that freezing began until freezing was complete? Did the entropy of the sample increase, decrease, or stay the same?

8. Predicting Outcomes
How would the quantity of the sample affect the time needed for the melting point test?

9. Predicting Outcomes
Would the quantity of the sample used to determine the melting point affect its outcome? (Hint: is melting point an extensive or intensive property?)

10. Interpreting Graphics

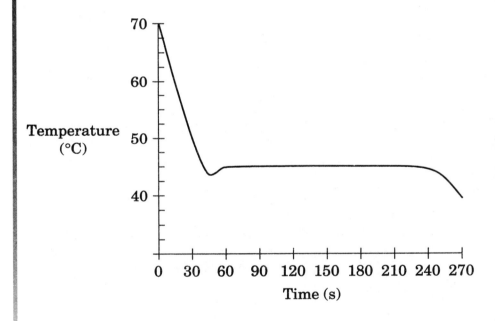

Examine the graph above, and compare it to your cooling curve. Would this sample of sodium thiosulfate pentahydrate be considered pure or impure? Sketch a line on the graph that represents your cooling data. If the curves are not identical, estimate the difference in melting points.

Extensions

1. Interpreting Graphics
Refer to the heating and cooling curves you plotted. For each portion of the curve, describe what happens to the energy and entropy of the substance.

2. Applying Ideas

In northern climates, freezing rain is a driving hazard. When this occurs, warm air from a defroster is blown against the windshield of an automobile in order to restore visibility. It would be convenient to have a system that automatically turned the defroster blower system on and off as needed. A thermostat embedded in the windshield to detect outside temperature could be used to perform this function.

a. At what temperature should the thermostat be set to turn on the hot-air blower?

b. At what temperature should the thermostat be set to turn off the hot-air blower?

3. Analyzing Methods

Will crystallization take place if no seed crystal is added? Why or why not?

4. Designing Experiments

Explain the purpose of a water bath. Why wasn't distilled water necessary?

5. Organizing Ideas

Which of the following word equations best represents the changes of phase taking place in the situations described in items 1 to 7 below? Place your answer in the space to the left of the numbered items.

a. solid + energy $\longrightarrow$ liquid

b. liquid $\longrightarrow$ solid + energy

c. solid + energy $\longrightarrow$ vapor

d. vapor $\longrightarrow$ solid + energy

_____ 1. ice melting at 0°C

_____ 2. water freezing at 0°C

_____ 3. a mixture of ice and water whose relative amounts remain unchanged

_____ 4. a particle escaping from a solid and becoming a vapor particle

_____ 5. solids, like camphor and naphthalene, evaporating

_____ 6. snow melting

_____ 7. snow forming

_____ 8. dry ice subliming

_____ 9. dry ice forming

Constructing a Cooling Curve—Melting Oils for Soap

March 14, 1995

Martha Li-Hsien
Director of Research
CheMystery Labs, Inc.
52 Fulton Street
Springfield, VA 22150

Dear Ms. Li-Hsien:

We are presently in the process of expanding our product line. In the past, we have focused on such items as air fresheners, carpet cleaners, and deodorizing cleansers. We are interested in producing a new line of fresh-smelling body soaps.

As the manufacturing process is developed, we must decide which fat to use in the production of our soap product. We are presented with three possibilities: tallow, vegetable shortening, and coconut oil. There are many factors we must consider, including the energy that the production will require. We would like you to investigate the energy required to melt these fats because soap production requires the fat to be liquid.

We are ready to pay you $175 000 if you give us the following information, which will assist us in deciding on which fat to use.

- a ranking of fats from least to greatest amount of energy required to change phase
- the approximate melting point of each fat
- a cooling curve for each fat
- data from which the cooling curve was graphed

Sincerely,

Martin Green

Martin Green
Vice President
Sparkle-Clean Industries

Memorandum

Date: March 15, 1995
To: Stacy Logan
From: Martha Li-Hsien

CheMystery Labs, Inc.
52 Fulton Street
Springfield, VA 22150

Sparkle-Clean Industries is the kind of client we need. Not only are they leading the market with their products, but they recently sold their dry-cleaning division and have plenty of capital to expand into related product lines. If we do well on this job, we could get more work from them as they test and develop other products as well. My plan is to come in under budget on this project as a way to attract their attention and encourage them to consider us for future work. So be careful about how much money you spend!

Because so much is at stake, I want to be kept informed about your work. Before you begin work, provide the following for me to review.
- a detailed, one-page plan for the procedure and all necessary data tables
- a detailed list of the equipment and materials you will need, along with the individual and total costs

After your preliminary work has been approved, you may begin your lab work. When your research team has completed the lab work, you should prepare a report in the form of a two-page letter to Martin Green at Sparkle-Clean Industries. Mr. Green outlined several requirements for the report in his letter. In addition to those, include the following items.
- a summary paragraph of how you solved this problem
- a discussion of the reliability of the data, including any possible sources of error
- a detailed invoice with costs for equipment, supplies, and services rendered

Required Precautions

Always wear goggles and a lab apron to provide protection for your eyes and clothing.

When you use a Bunsen burner, tie back loose clothing and hair. Do not heat glassware that is broken, chipped, or cracked. Use tongs whenever handling glassware or other equipment that has been heated. Do not directly heat the fats, as they are flammable. Instead, use a hot-water bath to heat them.

Do not touch any chemicals. If you get a chemical on your skin or clothing, wash it off at the sink while calling to your teacher.

Always wash your hands thoroughly when finished.

Spill Procedures/ Waste Disposal Methods

- If any of the fats spill, wipe up the spill with a paper towel.
- Dispose of the fats in the separate disposal containers designated by your teacher. You will need to re-heat the fats to transfer them to the disposal containers.
- The equipment used with the fats will need to be scrubbed well, using glassware brushes with plenty of soap and water.

References

The procedure for determining the heating/cooling curve is similar to a procedure you and your team recently completed. Because many fats such as tallow, shortening, and coconut oil are mixtures, not pure substances, it can be difficult to determine the precise melting point. Use at least 10.0 g of each fat so that you will be able to distinguish the melting point. Keep the fats in test tubes because otherwise they will float in the water and will not be evenly heated or cooled. First, heat the fats in a hot-water bath until they are slightly above their melting points (making a heating curve is not necessary). The cooling portion of the experiment can be carried out using a cold-water bath at about 15°C or colder.

MATERIALS for Sparkle-Clean Industries

REQUIRED ITEMS
(You must include all of these in your budget.)

Lab space/fume hood/utilities	15 000/day
Standard disposal fee	2 000/g
Balance	5 000
Beaker tongs	1 000

REAGENTS AND ADDITIONAL EQUIPMENT
(Include in your budget only what you will need.)

Samples of fats	500/g
400 mL beaker	2 000
100 mL graduated cylinder	1 000
Bunsen burner/related equipment	10 000
Buret clamp	1 000
Filter paper	500/piece
Forceps	500
Glass stirring rod	1 000
Hot plate	8 000
Litmus paper	1 000/piece
Ring stand/ring/wire gauze	2 000
Scissors	500
Spatula	500
Stopwatch	2 000
Test tube, large	1 000
Test-tube holder/rack	2 000
Thermistor probe — LEAP	2 000
Thermometer, nonmercury	2 000
Wash bottle	500
Watch glass	1 000
Weighing paper	500/piece

* No refunds on returned chemicals or unused equipment.

FINES

OSHA safety violation	2 000/incident

9-3 Heat of Fusion

Technique Builder

Situation

The City of Springfield is planning a new sports complex with a convertible arena. The floor of the arena will feature a system that will allow it to be flooded with water and frozen so that it that can be used for hockey or figure skating and then thawed and drained for events such as basketball and gymnastics. The city's recreation department has contacted your company for help in evaluating the cost-effectiveness of the new facility. They need to know how much energy is required to melt the ice so that they can determine how much it will cost each time they change the arena from an ice rink to a gymnasium.

Background

The freezing and melting points of water and ice are the same temperature: 0°C. The amount of heat energy that must be absorbed to melt 1 mol of solid water at 0°C is the same as the amount of heat energy released as 1 mol of water freezes. This quantity is called the molar heat of fusion. The units for this quantity are kilojoules per mole (kJ/mol). With this value, you can predict the amount of heat energy required to melt any measured quantity of ice.

Problem

To help the city determine the amount of energy needed to melt an ice rink, you must first determine the molar heat of fusion by doing the following.

- Measure the volume and temperature of some warm water in a calorimeter.
- Add some ice to the warm water, and measure the temperature change and the change in volume after melting is complete.
- Determine how much ice was added.
- Calculate the amount of heat released as the warm water cooled, using the equation for specific heat capacity.
- Calculate the molar heat of fusion of the ice.

Objectives

Measure the volume of a sample of water.

Calculate the mass of water given the volume, and then determine the number of moles present.

Use a calorimeter to measure the changes in temperature that occur as ice melts.

Relate temperature changes to enthalpy changes.

Calculate the molar heat of fusion of ice.

Materials

- Ice
- 400 mL beakers, 2
- 100 mL graduated cylinder
- Balance
- Beaker tongs
- Calorimeter or two plastic foam cups
- Glass stirring rod
- Ring stand

Bunsen burner option

- Bunsen burner and related equipment
- Ring clamp
- Wire gauze with ceramic center

Hot plate option

- Hot plate

LEAP System option

- LEAP System
- LEAP thermistor probe

Thermometer option

- Thermometer, non-mercury
- Thermometer clamp

Safety

Always wear goggles and a lab apron to provide protection for your eyes and clothing. If you get a chemical in your eyes, immediately flush it out at the eyewash station while calling to your teacher. Know the locations of the emergency lab shower and eyewash station and how to use them.

When you use a Bunsen burner, confine any long hair and loose clothing. Do not heat glassware that is broken, chipped, or cracked. Use tongs or a hot mitt to handle heated glassware and other equipment because hot glassware does not look hot. If your clothing catches on fire, WALK to the emergency lab shower and use it to put out the fire.

Always clean up the lab and all equipment after use, and dispose of substances according to proper disposal methods. Wash your hands thoroughly before you leave the lab after all lab work is finished.

Procedure

1. **Organizing Data**

   Prepare a data table in your lab notebook with four columns and five rows. In the first row, label the boxes *Measurement, Trial 1, Trial 2,* and *Trial 3.* In the first column, label the second through fifth boxes *Init. H_2O vol. (mL), Init. H_2O temp. (°C), Final H_2O vol. (mL),* and *Final H_2O Temp. (°C).*

2. If a calorimeter that has already been made is not available, construct one from two plastic foam cups. Trim the lip of one cup and use that cup for the top of the calorimeter. Make a small hole in the base of this cup (the top of the calorimeter) so that a thermometer or thermistor can be inserted, as shown in the illustration below. Place the calorimeter in one of the beakers to help prevent it from tipping over.

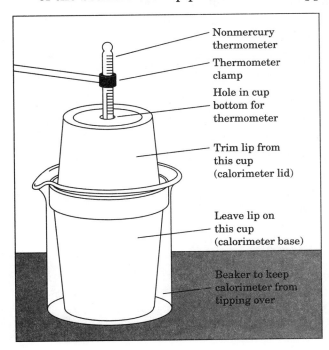

Technique

3. Pour about 350 mL of water into a 400 mL beaker.

4. If you use a Bunsen burner, attach a ring clamp to a ring stand. Place the wire gauze with ceramic center and the beaker on top of the ring clamp. If you use a hot plate, place the beaker directly on the hot plate.

5. Heat the beaker until the water has reached a temperature of about 45°C. Check the temperature every few minutes with a thermometer or LEAP System with thermistor probe. Do not measure the temperature too close to the surface of the water or too close to the bottom of the beaker.

6. Using beaker tongs or a hot mitt, pour about 100.0 mL of the heated water from the beaker into a graduated cylinder. Measure the volume to the nearest 0.1 mL, and record it in your data table as the initial volume for *Trial 1*.

7. Pour the water into the calorimeter cup, and measure the temperature to the nearest 0.1°C. Record this temperature as the initial temperature for *Trial 1*.

8. Holding the ice cubes with the beaker tongs, shake any excess water off them. Quickly place the ice in the warm water in the calorimeter, and stir the ice water with a glass stirring rod. Place the lid on the calorimeter, and gently insert the thermometer into the hole on the top of the calorimeter lid. Attach the thermometer to the ring stand using the thermometer clamp. Measure the temperature every minute. As soon as the temperature is below 3°C, measure the lowest temperature reached in the next 5 min to the nearest 0.1°C. Record this value in your data table as the final temperature for Trial 1. *Never use a thermometer to stir anything. It will break easily. The glass wall surrounding the bulb is very thin to provide quick and accurate temperature readings.*

9. Remove the calorimeter lid. Using the beaker tongs, remove any unmelted ice. Allow any water on the ice and tongs to drip back into the calorimeter.

10. Using the graduated cylinder, measure the volume of water remaining in the calorimeter to the nearest 0.1 mL. Record this value in your data table as the final volume for *Trial 1*. Because the final total volume will probably be larger than 100.0 mL, it may be necessary to measure the volume of about half of the water in the calorimeter, empty the graduated cylinder, and then measure the volume of the rest of the water.

11. Repeat steps **5–10** twice, recording the results in your data table in the columns for *Trial 2* and *Trial 3*.

Cleanup and Disposal

12. All water and ice may be washed down the drain. Clean up your equipment and lab station. Always wash your hands thoroughly before you leave the lab after cleaning up the lab area and equipment.

Analysis and Interpretation

1. **Organizing Data**
   Calculate the volume of water that came from the melted ice for each trial.

2. **Analyzing Data**
   Determine the mass of ice that melted for each trial. (Hint: the density of water is 1.00 g/mL from 0°C to 30°C.)

3. **Analyzing Data**
   Using a periodic table, determine the molar mass of water, and calculate how many moles of water were contained in the ice that melted for each trial.

4. **Analyzing Data**
   Determine the mass of the warm water for each trial. (Hint: the density of water is 0.992 g/mL at 40°C, 0.990 g/mL at 45°C, 0.988 g/mL at 50°C, and 0.986 g/mL at 55°C.)

5. **Organizing Data**
   Calculate the change in temperature, Δt, of the warm water for each trial.

6. **Organizing Data**
   Calculate the change in temperature, Δt, of the water that was melted from the ice for each trial. (Hint: assume that the temperature was 0°C initially.)

Conclusions

7. **Organizing Ideas**
   Write a valid equation relating the following three quantities: the energy absorbed by the ice as it melted, the energy absorbed by the melted water as it warmed, and the energy released by the warm water as it cooled. (Hint: assume that the calorimeter is a perfect insulator. Assume that the ice was at 0°C initially.) Explain the scientific principle that allows you to link these quantities in an equation.

8. **Applying Ideas**
   Identify which two of the quantities in the equation from item **7** can be calculated from information you already have. (Hint: apply the equation for specific heat capacity. Assume that the ice was at 0°C initially.)

9. **Applying Ideas**
   Set up three different sets of calculations, one for each trial, to determine the values of the quantities identified in item **8**. (Hint: the specific heat capacity of water is 4.184 J/g·°C.)

10. **Inferring Conclusions**
    How much heat energy was absorbed by the ice in each of the trials?

11. **Organizing Conclusions**
    On average, how much heat energy is required to melt 1.00 g of ice?

12. **Organizing Conclusions**
    Based on your measurements, what is the molar heat of fusion for water?

13. Applying Conclusions

The rink that the city is proposing would be roughly 25 m ×
60 m, and the ice would be 2.5 cm thick. If the density of ice is
0.917 g/mL, how much energy would be required to melt this
ice?

Extensions

1. Evaluating Conclusions

Compare your values for the molar heat of fusion of water to
values that can be found in a handbook such as the *CRC
Handbook of Chemistry and Physics* or *Lange's Handbook of
Chemistry*. (You may need to convert the values given in the
handbook from calories to joules in order to use them. The conversion factor is 1 cal = 4.184 J.) Calculate your percent error.

2. Evaluating Conclusions

Calibrate the calorimeter, following the procedure given in the
first part of the Technique section in Exploration 9-1. Apply
the results to your calculations for this Exploration. Compare
these results with those you determined in item **10** of the Conclusions section.

3. Designing Experiments

What possible sources of error can you identify with this procedure? If you can think of ways to eliminate them, ask your
teacher to approve your plan, and try it out.

4. Applying Ideas

A hundred years ago in the United States, vegetables stored
for the winter were kept in a "cold cellar." This was an un-
heated room beneath a house. To prevent these food supplies
from freezing during severe winter weather, large barrels of
water were placed in the rooms. Explain why this kept the
temperature from falling below 0°C.

Heat of Solution

Objectives

Measure **tempera-ture changes in a calorimeter as a solid solute is dissolved.**

Relate **temperature changes in a calorimeter to changes in heat energy.**

Determine **the amount of heat released in joules for each gram of $Na_2S_2O_3 \cdot 5H_2O$ or $NaC_2H_3O_2$ that dissolves.**

Evaluate **the use of $Na_2S_2O_3 \cdot 5H_2O$ and $NaC_2H_3O_2$ in a heating pad.**

Situation

In the summer of 1986, a commercial company began manufacturing an unusual heating pad. It is filled with a liquid solution that exists in a supercooled state at normal temperatures. To use it, a metal disc on the heating pad is pressed and flexed rapidly, causing vibrations that disturb the solution and induce crystallization, which releases a large amount of heat. The heating pad can be reused by placing it in boiling water to remelt the crystals and then slowly cooling it to room temperature. You have been hired by this firm to explore new product possibilities using a solution of sodium thiosulfate, $Na_2S_2O_3 \cdot 5H_2O$, instead of sodium acetate, $NaC_2H_3O_2$. An important part of this feasibility study is to determine the amount of heat given off per gram of substance, the amount of the substance needed for a heating pad, and the total cost of the substance needed.

Background

The solution in this heating pad is made by dissolving a salt in water. It is a supersaturated solution; that is, it holds more dissolved salt than is usually possible at a particular temperature. The solution is metastable, so disturbing it causes the ions to become ordered enough to crystallize out of solution, releasing heat energy. The same amount of heat energy is involved when crystals dissolve in water. The bombarding of the crystal lattice by water molecules causes it to break apart. Then these free ions break the hydrogen bonds between water molecules. The water molecules surround the ions, attracted by their charge, and hydrate them. As these interactions take place, energy is released. The sum of the enthalpies of these processes is the heat of solution, ΔH_{sol}. If the hydration step, which involves bond formation, releases more heat than the bond-breaking step, the process is exothermic and ΔH_{sol} is negative. Otherwise, dissolving is endothermic and ΔH_{sol} is positive.

Problem

To compare the cost effectiveness of using $Na_2S_2O_3 \cdot 5H_2O$ instead of $NaC_2H_3O_2$ in a heating pad, you must do the following.

• Construct a calorimeter and dissolve each salt in it.

• Calculate Δt and ΔH for the water in which the salts dissolve.

• Determine the supplier's price for each salt.

• Calculate ΔH per gram and the cost-effectiveness for each salt.

• Analyze the precision of your results.

Safety

Always wear goggles and a lab apron to provide protection for your eyes and clothing. If you get a chemical in your eyes, immediately flush it out at the eyewash station while calling to your teacher. Know the locations of the emergency lab shower and eyewash station and how to use them.

Do not touch any chemicals. If you get a chemical on your skin or clothing, wash it off at the sink while calling to your teacher. Carefully read the labels and follow the directions on all containers of chemicals that you use. Do not taste any chemicals or items used in the laboratory. Never return leftovers to their original containers; take only small amounts to avoid wasting supplies.

Always clean up the lab and all equipment after use, and dispose of substances according to proper disposal methods. Wash your hands thoroughly before you leave the lab after all lab work is finished.

Preparation

1. **Organizing Data**

   Prepare a data table with five columns and five rows. Label the boxes in the first row *Measurement, Trial 1—$Na_2S_2O_3 \cdot 5H_2O$, Trial 2—$Na_2S_2O_3 \cdot 5H_2O$, Trial 3—$NaC_2H_3O_2$,* and *Trial 4—$NaC_2H_3O_2$*. Label the second through fifth boxes in the first column *Mass of solute (g), Volume of cold H_2O (mL), Initial H_2O temp. (°C),* and *Final H_2O temp. (°C)*.

2. Prepare a cold-water bath. Fill a small, plastic washtub with ice. Fill a 600 mL beaker three-fourths full of distilled water. Make a hole in the ice large enough for the beaker. Insert the beaker, and pack the ice around it up to the level of the water.

3. Prepare the calorimeter. Cut a square of corrugated cardboard slightly larger than the top of a 4 oz plastic foam cup. Make a hole in the center of the cardboard piece with a pencil. Make a second hole in the cardboard piece less than 1.0 cm away from the hole in the center. Insert a piece of wire through the hole, and bend each end to make 1.0 cm loops, as shown in the illustration on the next page. Insert a thermometer or LEAP thermistor probe into the center hole and set the entire assembly aside until step **10**.

4. On a piece of weighing paper, measure the mass of approximately 15 g of $Na_2S_2O_3 \cdot 5H_2O$ to the nearest 0.01 g. Record the mass in your data table.

5. Completely transfer the $Na_2S_2O_3 \cdot 5H_2O$ to a small Pyrex test tube.

6. Label the tube *Trial 1,* and set it in a test-tube rack.

7. Repeat steps **3–5** for the following.
   a. 15 g of $Na_2S_2O_3 \cdot 5H_2O$; label the test tube *Trial 2*
   b. 15 g of $NaC_2H_3O_2$; label the test tube *Trial 3*
   c. 15 g of $NaC_2H_3O_2$; label the test tube *Trial 4*

Materials

- Ice
- $NaC_2H_3O_2$
- $Na_2S_2O_3 \cdot 5H_2O$
- 600 mL beaker
- 100 mL graduated cylinder
- Balance, centigram
- Corrugated cardboard or lid for cup
- Pencil with point
- Plastic washtub
- Plastic foam cup, small
- Test tubes, small, Pyrex, 4
- Test-tube rack
- Wire, 10–15 cm

LEAP System option
- LEAP System
- LEAP thermistor probe

Thermometer option
- Thermometer, non-mercury

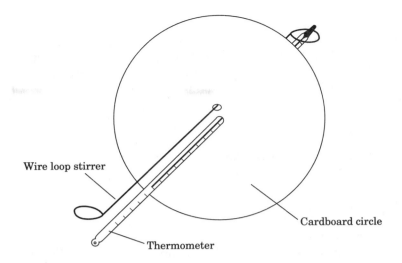

Wire loop stirrer

Cardboard circle

Thermometer

Technique

8. Pour approximately 75 mL of the cold water prepared in **step 2** into a 100 mL graduated cylinder. Record the volume to the nearest 0.1 mL.

9. Pour the cold water from the graduated cylinder into the plastic foam cup.

10. Put the thermometer or thermistor assembly into the cup. The bulb of the thermometer should be completely covered by the water but must not touch the bottom of the cup.

11. Record the temperature of the cold water to the nearest 1.0°C if using a thermometer or to the nearest 0.1°C if using the LEAP System with thermistor probe.

12. Lift the cardboard slightly and dump the contents of the *Trial 1* test tube into the plastic foam cup all at once.

13. Gently move the stirring wire up and down inside the cup to disperse the heat. Allow the solid to dissolve completely.

14. Stir continuously until the temperature of the water peaks.

15. Record the highest temperature reached by the water to the nearest 1.0°C if using a thermometer or to the nearest 0.1°C if using the LEAP System with thermistor probe.

16. Remove, rinse, and dry the thermometer or thermistor assembly.

17. Pour the solution into the designated waste container.

18. Rinse and dry the cup.

19. Repeat steps **8** through **18** for *Trial 2, Trial 3,* and *Trial 4.*

Cleanup and Disposal

20. Both substances, $Na_2S_2O_3 \cdot 5H_2O$ and $NaC_2H_3O_2$, and their solutions, should be placed in their own designated disposal container. Remember to wash your hands thoroughly after cleaning up your lab area and equipment.

Analysis and Interpretation

1. Organizing Data
Determine the change in temperature, Δt, of the cold water for each trial.

2. **Organizing Information**
   Calculate the heat energy in joules absorbed by the cold water for each trial, using the specific heat capacity equation found on page 316 of the textbook. Assume the density of water is 1.00 g/mL and the specific heat capacity of water is 4.180 J/g•°C.

3. **Organizing Data**
   Calculate the amount of heat energy released per gram of $Na_2S_2O_3•5H_2O$ for *Trial 1* and *Trial 2*.

4. **Organizing Data**
   Calculate the amount of heat energy released per gram of $NaC_2H_3O_2$ for *Trial 3* and *Trial 4*.

5. **Analyzing Information**
   The accepted value for the heat of solution for $Na_2S_2O_3•5H_2O$ is 200.0 kJ/g. Calculate your percent error.

6. **Analyzing Information**
   Calculate the average of *Trial 1* and *Trial 2*, and the average of *Trial 3* and *Trial 4*. Calculate the percent difference between each trial and the average. Comment on your precision.

7. **Analyzing Information**
   Locate a supplier and find out the cost for $Na_2S_2O_3•5H_2O$ and for $NaC_2H_3O_2$. Calculate the cost per gram for each, and comment on which substance would be the most cost-effective as the primary ingredient of a heat pack.

8. **Evaluating Methods**
   Why was a plastic foam cup instead of a beaker used as a reaction vessel?

9. **Evaluating Methods**
   How could the apparatus for measuring the temperature change of the water be improved?

Conclusions

10. **Using Information**
    What is the source of the heat energy that is released by the $Na_2S_2O_3•5H_2O$ and absorbed by the water?

11. **Inferring Conclusions**
    What would be the ΔH of a reaction in which $NaC_2H_3O_2$ crystallizes out of solution?

Extensions

1. **Inferring Conclusions**
   Why is the fact that the temperature of a phase change remains constant for a pure substance crucial to the development and use of this heating pad?

2. **Relating Ideas**
   When ionic substances dissolve in water, the crystal lattice is broken, and the ions are immediately surrounded by water molecules in a process called hydration. Hydration is exothermic. Breaking the lattice absorbs heat energy. ΔH_{sol} is the net energy resulting from overcoming the lattice energy and releasing the energy of hydration. Draw a diagram relating ΔH_{lat}, ΔH_{hyd}, and ΔH_{sol}.

10-1 *Technique Builder*

Gas Pressure-Volume Relationship

Objectives

Measure the volume of a gas as pressure is increased.

Calculate the pressure on a gas given mass, acceleration due to gravity, and area.

Graph the pressure-volume data in two ways.

Express the pressure-volume relationship mathematically.

Calculate the amount of gas contained in a tank at different pressures.

Situation

Your company has been contacted to do some research and development work for the Valvopress Company. Valvopress makes valves for a variety of compressed-air tools, and they hope to use this expertise to expand into the scuba-tank market. They need you to help them decide which of several different valves they make would be the best one to install in the new tank. Valvopress has designed a prototype tank that has a volume of 10.5 L, but they need to know, for each of the possible valves, what volume of air at 1.0 atm of pressure could be compressed into a fully pressurized tank.

Background

In Chapter 10, the relationship between pressure and volume of a sample of gas at constant temperature, Boyle's law, was described. You will make measurements to test Boyle's law using a sample of air enclosed in the tube of a Boyle's law apparatus. After measuring the volume at normal pressure, you will measure how the volume changes as pressure increases.

As was discussed in Chapter 10, pressure is equal to the amount of force exerted on an area. You will increase the pressure by stacking objects on top of a plunger. Because of gravity, the objects will exert a force (in N) equal to 9.801 m/s^2 times the value of their mass (in kg).

Remember that normal pressure at sea level is defined to be exactly 1 atm. Unless your teacher instructs you otherwise, use this value as the normal pressure in your work. Remember: 1.00 atm = 29.92 in of Hg = 760 mm of Hg = 1.013×10^5 Pa = 101.3 kPa.

Problem

To evaluate the different valves, your company needs to do the following.
- Vary the pressure on a fixed amount of air and measure the volume.
- Plot graphs of pressure versus volume and pressure versus the inverse of volume.
- From the graphs, determine how pressure and volume are related.
- From the relationship between pressure and volume, calculate what volume of air at normal pressure is equivalent to a tankful of pressurized air for each of the valves.

Safety

Always wear goggles and a lab apron to provide protection for your eyes and clothing. If you get a chemical in your eyes, immediately flush it out at the eyewash station while calling to your teacher. Know the locations of the emergency lab shower and eyewash and how to use them.

Always clean up the lab and all equipment after use, and dispose of substances according to proper disposal methods. Wash your hands thoroughly before you leave the lab after all lab work is finished.

Preparation

1. Organizing Data

Prepare a data table with six columns and five rows. In the first through fifth rows of the first column, write the following labels: *Masses added, Mass (g), Trial 1 vol. (cm³), Trial 2 vol. (cm³),* and *Trial 3 vol. (cm³).* In the second through sixth columns of the first row, write these labels: *0 mass, Mass 1, Mass 2, Mass 3,* and *Mass 4.* You will also need a space to record the diameter of the syringe barrel below the data table.

2. The syringe or your Boyle's law apparatus is marked in units of cubic centimeters, abbreviated cc or cm³. Volume in cubic centimeters is the same as volume in milliliters.

3. Using removable Post-It notes, label the objects that will be stacked on the plunger of the syringe *1* through *4.*

Technique

4. Using a balance, measure the mass of each of the four objects you will use as masses, with their labels attached. Record each mass in the appropriate space in the data table. Under the column labeled *0 mass,* fill in a zero.

5. Using a metric ruler, record the diameter of the syringe or Boyle's law apparatus to the nearest millimeter, and record it in your lab notebook.

6. Set up the syringe as your teacher instructs so that you will be able to place the masses on the plunger without the apparatus falling over. If you are using a Boyle's law apparatus, this step is not necessary.

7. If you are using a Boyle's law apparatus, trap about 35 cm³ of air in the syringe of the Boyle's law apparatus. If you are using a sealed syringe, trap about 9 cm³ of air in the syringe. For either piece of equipment, pull the plunger all of the way out of the syringe. Insert a piece of carpet thread into the barrel with the end of the thread hanging out to make a small space for air to escape, as shown in the illustration on the following page. Push the plunger back into the syringe and position the head at the appropriate location, either 35 cm³ for a Boyle's law apparatus or 9 cm³ for a sealed syringe.

Materials

- Balance
- Carpet thread
- Removable Post-It notes
- Ruler, metric
- Sealed syringe or Boyle's law apparatus
- Stackable objects of roughly equal mass, such as books or blocks of wood (500–550 g each), 4

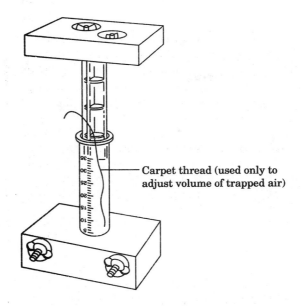

Carpet thread (used only to adjust volume of trapped air)

8. While holding the plunger in place, carefully pull out the thread. Twist the plunger several times to allow the head to overcome any frictional forces. Read the volume to the nearest 0.1 cm³, and record this value in your data table as the initial volume for zero mass.

9. Place the first of the masses on the plunger. Your partner should stand by to ensure that the masses do not fall. Give the plunger several twists to reduce frictional forces. When the plunger comes to rest, read and record the volume to the nearest 0.1 cm³ in the row for *Trial 1*.

10. Carefully position the second mass on top of the first. Your partner should stand by to ensure that the masses do not slip off as you twist the plunger. Allow the plunger to come to rest before you read and record the volume.

11. Follow the procedure described in step **10** as you add the third and fourth masses. Record the volume after each addition.

12. Remove all four masses and repeat the process two more times, starting with step **9**, recording the results in the rows for *Trial 2* and *Trial 3*.

Cleanup and Disposal

13. Clean all apparatus and your lab station. Remember to wash your hands.

Analysis and Interpretation

1. Organizing Data
By adding the masses you measured, calculate the total mass that was on the plunger for each of the volumes measured (zero to four masses).

2. Analyzing Data
Determine the cross-sectional area of the plunger using the diameter recorded in the data table. Then convert your answer into units of square meters, m².

$$\left(\text{Hint: for a circle, area } = \pi\left(\frac{\text{diameter}}{2}\right)^2\right)$$

3. **Analyzing Data**
   The force in newtons that masses exert due to gravity is equal to their mass in kilograms times the value of their acceleration due to gravity. For objects near the Earth's surface, the acceleration due to gravity, g, has a value of 9.801 m/s². Calculate the force exerted by the masses on the plunger for each of the volumes measured.

   $$F = mg = m \times 9.801 \text{ m/s}^2$$

4. **Organizing Data**
   Pressure is the force exerted divided by the area over which the force is exerted. If the force is in units of newtons and the area is in units of square meters, the pressure calculated will be in units of pascals (Pa). Calculate the pressure exerted by the masses for each of the volumes measured.

5. **Resolving Discrepancies**
   Why would it be incorrect to say that there is no pressure on the trapped air when no masses are on the plunger? What is the total pressure exerted on the sample before any masses have been added to the plunger? (Hint: it is not equal to zero.)

6. **Analyzing Data**
   Using your answers to items **4** and **5**, calculate the total pressure for each stage of the procedure.

7. **Organizing Data**
   Calculate the average volume you measured in the three trials for each of the five volumes measured.

8. **Analyzing Data**
   Examine your data table. What conclusions can you draw about the relationship of pressure to volume?

9. **Interpreting Graphics**
   Make a graph of pressure versus volume. Plot pressure on the horizontal axis and average volume on the vertical axis. Draw a smooth curve through the points.

10. **Organizing Data**
    Calculate 1/V, the inverse of each of the average volumes. For example, if the average volume is 26.5 cm³, then 1/V =

    $$\frac{1}{26.5 \text{ cm}^3} = 0.0377 \text{ cm}^{-3}.$$

11. **Interpreting Graphics**
    Make a graph of pressure versus 1/V. Plot pressure on the horizontal axis and 1/V on the vertical axis. Draw the best-fit line through the points.

12. **Interpreting Graphics**
    A straight line has an equation of the form $y = mx + b$. From your graph, what are the values of m and b if 1/V and P are substituted for y and x, respectively? (Hint: if you have a graphics calculator, use the $\boxed{\text{STAT}}$ mode to enter your data and make a linear regression equation using the LinReg function from the STAT menu.)

13. **Analyzing Methods**
    Explain how you can be certain that neither the temperature nor the number of moles of gas changed during the course of the procedure.

14. Analyzing Methods

What would have happened if there had been no carpet thread in step **7** and you tried to adjust the plunger?

Conclusions

15. Inferring Conclusions

Rearrange your equation from item **12** so that P and V are on the same side of the equation. Substitute the symbol k for the constant term. (Hint: the approximation of very small amounts as being equal to zero may be necessary.)

16. Evaluating Conclusions

Consider the value of k that was found for the sample of gas in the procedure. Can you use the same value of k for pressure and volume calculations for the scuba tank? (Hint: consider what variables were held constant during the procedure.)

17. Applying Conclusions

Using the equation from item **15,** how can you relate these four quantities: the pressure in the scuba tank, the volume of the air in the tank at that pressure, normal atmospheric pressure, and the volume of the air in the scuba tank if it was at normal atmospheric pressure.

18. Applying Ideas

Using your answer to item **17,** calculate the volume of air at normal atmospheric pressure that could be contained in a scuba tank at the maximum pressure for each of the valves listed in the table shown below.

Valve Serial Number	Top Pressure
PDE-57	112 atm
RRM-82	168 atm
RMD-332	219 atm
ZRH-61	240 atm
XJP-27	253 atm

19. Applying Ideas

For every 1000 L of air at normal atmospheric pressure, a diver can spend, on average, 26 min underwater at a depth of 20 ft. Using your answer to item **18,** calculate how long you could stay underwater with tanks made using each of the valves.

20. Evaluating Viewpoints

The unit costs for the different possible components of the scuba tank are given in the following table. What will be the unit cost of the most efficient scuba tank, in terms of volume of gas at normal atmospheric pressure compared to cost?

Item	Cost
Filling tank with air	$ 1.50
External pressure meter	$ 20.00
Tank	$ 50.00
PDE-57 valve	$ 37.50
RRM-82 valve	$ 40.00
RMD-332 valve	$ 45.00
ZRH-61 valve	$ 70.00
XJP-27 valve	$105.00

Extensions

1. **Connecting Ideas**
   Explain how the relationship you observed between the pressure and volume of a gas is consistent with the kinetic molecular theory of gases.

2. **Evaluating Methods**
   The value of the pressure that you applied to the gas in the form of masses on the plunger of the syringe is referred to as the value of the pressure of the gas itself. Explain why this is logical.

3. **Designing Experiments**
   What possible sources of error can you identify in your procedure? If you can think of ways to eliminate them, ask your teacher to approve your suggestions, and run the procedure again.

4. **Applying Conclusions**
   Imagine a gas confined in an unbreakable container. In theory, the pressure on the gas could be increased gradually until it approached infinity. What happens to the volume of the gas as the pressure becomes infinitely high?

5. **Research and Communication**
   Find out about the original experiment that Robert Boyle performed to establish the relationship you just explored. Write a short paper describing his experimental setup and procedure. Include a sketch of the equipment he used, and compare it to the Boyle's law apparatus you used.

Gas Temperature-Volume Relationship— Balloon Flight

March 15, 1995

Ms. Sandra Fernandez
Director of Development
CheMystery Labs, Inc.
52 Fulton Street
Springfield, VA 22150

Dear Ms. Fernandez:

Our corporation, Barotherm, Inc. is developing a new type of hot-air balloon. Instead of using gas heaters, we plan to use electric heaters powered by new flexible solar panels on the balloon's top. In this way, the balloon can be kept sealed.

We are also creating an "autopilot" system to control the balloon's flight using temperature sensors sewn into the fabric of the balloon. In order to program the system's computer, we need you to determine the internal temperature necessary to keep the balloon aloft.

The balloon can hold as much as 2.6×10^3 m³ of air, but it may not be necessary to fully inflate the balloon to achieve the proper amount of lift. Before heating, the balloon will be sealed at 15°C with 2.0×10^3 m³ of air inside. The mass of the balloon's equipment and payload will be 480 kg.

To prepare for these calculations, experimentally test the relationship between temperature and volume for a sample of air. Any deviations from ideal gas behavior should be incorporated into your calculations.

We are willing to pay $100 000 for this work, provided it can be completed in a timely fashion so that we can begin testing the prototype model as soon as possible.

Sincerely

Vitra Muhmedi

Vitra Muhmedi

Memorandum

CheMystery Labs, Inc.
52 Fulton Street
Springfield, VA 22150

Date: March 16, 1995
To: Leah Quarters
From: Sandra Fernandez

As you know, according to Charles's law, an ideal gas sample will have a volume proportional to the temperature in kelvins. But real gases such as air can show some deviations from this ideal behavior. Thus, it is necessary for us to test samples of air ourselves.

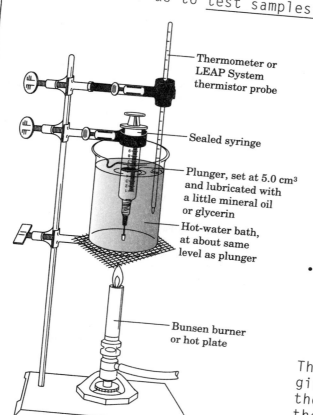

- Thermometer or LEAP System thermistor probe
- Sealed syringe
- Plunger, set at 5.0 cm³ and lubricated with a little mineral oil or glycerin
- Hot-water bath, at about same level as plunger
- Bunsen burner or hot plate

I propose using a setup like the one sketched here. Before you begin the work, I need the following.

- detailed one-page plan and all necessary data tables
- detailed list of all equipment and materials you will need, along with individual and total costs
- well-organized calculations for the necessary balloon temperature, assuming air is an ideal gas

The balloon will begin to float when the combined mass of the payload, the balloon, and the heated air inside it is less than the mass of the air displaced by the balloon's volume. Some necessary information is given in the References section.

When you have completed your work, prepare a report in the form of a two-page letter to Ms. Muhmedi at Barotherm. Be sure to cover the following points.

- temperature and volume necessary for the balloon to float, along with calculations
- brief description of the procedure
- detailed and organized data section
- organized analysis section explaining and graphing the data, as well as any possible sources of error
- detailed invoice for services and expenses

Required Precautions

 Always wear goggles and a lab apron to provide protection for your eyes and clothing.

 Tie back loose clothing and hair. Do not heat glassware that is broken, chipped, or cracked. Use tongs whenever handling glassware or other equipment that has been heated because hot glassware does not always look hot.

 If you will be using a thermometer, be sure that you are using a thermometer clamp to hold it in place or that the thermometer has already been inserted into a split one-hole stopper. Clamp the stopper, not the thermometer itself.

 Always wash your hands thoroughly when finished.

Spill Procedures/ Waste Disposal Methods

- Water from the hot-water bath may be poured down the drain.
- Be sure to thoroughly clean the plunger of the syringe with soap and water.

References

Charles's law, which relates temperature and volume for gases, is discussed in the textbook on pages 368–371.

On the ground, the density of air at 15°C and normal pressure is 1.22 g/L. At the balloon's flying altitude of 500 m, the temperature drops slightly, and the density of air is 1.17 g/L.

MATERIALS for Barotherm, Inc. balloon project
REQUIRED ITEMS
(You must include all of these in your budget.)

Lab space/fume hood/utilities	15 000/day
Standard disposal fee	2 000/g of product
Beaker tongs	2 000

REAGENTS AND ADDITIONAL EQUIPMENT
(Include in your budget only what you'll need.)

Lubricant (mineral oil or glycerin)	1 000/mL
400 mL breaker	2 000
100 mL graduated cylinder	1 000
10 cm³ sealed syringe	5 000
Bunsen burner/related equipment	10 000
Clamps	1 000
Filter paper	500/piece
Glass funnel	1 000
Glass stirring rod	1 000
Hot plate	8 000
Ring stand/ring/wire gauze	2 000
Six test tubes/holder/rack	2 000
Thermistor probe—LEAP	2 000
Thermometer, nonmercury, in one-hole stopper	2 000
Wash bottle	500
Weighing paper	500/piece

* No refunds on returned chemicals or unused equipment.

FINES

OSHA safety violation	2 000/incident

10-2 Masses of Equal Volumes of Gases

Technique Builder

Situation

Compressed Gases, Inc. provides gases to customers in heavy steel tanks. For safety's sake, tanks are filled and refilled with only one gas, and a different color is used for tanks of each gas. At the factory of one customer, an overenthusiastic painter painted several different tanks the same color. The customer believes that the tanks contained carbon dioxide, CO_2, and methane, CH_4, but isn't sure. The customer also uses hydrogen, H_2, and oxygen, O_2. In their pure form, all of these gases are odorless and colorless, so they are hard to distinguish.

Background

Avogadro's principle states that equal volumes of gas, measured at the same temperature and pressure, contain the same number of molecules. Thus, if you measure the masses of these equal volumes, you can be sure that you have measured the masses of equal numbers of molecules. In other words, the ratio is the same as the mass ratio of a molecule of CO_2 to a molecule of CH_4.

However, when the mass of an object is measured in a fluid such as air, the fluid buoys up the object, making its mass appear to be less. The difference between the apparent and actual mass is equal to the mass of the air that is displaced by the object's volume. For most solid and liquid objects, the difference is small. But for a gas, its mass is close to the mass of the air it displaces. To find the actual mass, the mass of the air displaced by the gas must be added to the apparent mass of the gas. Suppose the volume of a gas is exactly 1.00 L at 20°C and 760 mm Hg. At this temperature and pressure, the density of air is 1.2 g/L. The apparent mass of the gas will be 1.2 g less than if it was measured in a vacuum. To obtain the actual mass of the gas, 1.2 g must be added to its apparent mass.

Problem

The staff at Compressed Gases, Inc. determined the ratio of the masses of the two gases as 2.0 : 1.0. In order to verify the identities of the gases, you must do the following.

- Experimentally verify this mass ratio at constant temperature and pressure using known samples of pure CO_2 and pure CH_4.
- Compare your results to those of Compressed Gases, Inc. to decide whether the tanks are likely to contain CO_2 and CH_4.
- If so, explain which tank contains which gas.
- If not, explain which gases the tanks do contain.

Objectives

Measure the masses of equal volumes of methane and carbon dioxide under the same conditions of temperature and pressure.

Measure the volume of the gas in the plastic bag by water displacement.

Determine the actual masses of the gases by eliminating the effect of buoyancy.

Calculate the ratio of the two masses.

Apply Avogadro's principle to determine the identity of gases by determining the ratios of their molar masses.

Materials

- Dry ice
- Methane gas (from gas jet)
- 2 L glass bottle
- 250 mL or 500 mL graduated cylinder
- 1 L plastic bag
- Balance
- Crucible tongs
- Medicine dropper inserted in a rubber stopper
- Pinch clamp
- Plastic straw
- Pneumatic trough
- Rubber band
- Rubber tubing, 50 cm

Safety

Always wear goggles and a lab apron to provide protection for your eyes and clothing. If you get a chemical in your eyes, immediately flush it out at the eyewash station while calling to your teacher. Know the locations of the emergency lab shower and eyewash and how to use them.

Do not touch any chemicals. If you get a chemical on your skin or clothing, wash it off at the sink while calling to your teacher. Make sure you carefully read the labels and follow the directions on all containers of chemicals that you use. Do not taste any chemicals or items used in the laboratory. Never return leftovers to their original containers; take only small amounts to avoid wasting supplies.

Methane is flammable, and mixtures of methane and air can explode if ignited. Make sure there are no flames anywhere in the room while anyone is working with methane. Do not turn any electric switches on or off while anyone is working with methane. When you finish filling the bag, check twice to be sure that your gas valve and the next closest one are turned off.

Never handle dry ice with your hands; it is cold enough to cause burns. Use tongs to pick it up, or wear well-insulated cloth gloves when you touch it. Wrap the medicine dropper in a layer of cloth and wear disposable plastic gloves when fitting the medicine dropper into the tubing.

Carbon dioxide is an asphyxiant. If it is stored in a container such as an ice chest, do not put your head near it.

Always clean up the lab and all equipment after use, and dispose of substances according to proper disposal methods. Wash your hands thoroughly before you leave the lab after all lab work is finished.

Preparation

1. Organizing Data

Prepare a data table with spaces for the following: the room temperature and pressure in the lab, the volume of the bag, the mass of the bag assembly, and the mass of the bag assembly when filled with CO_2 and with CH_4.

Technique

2. The plastic bag you use should be dry and free of holes. Fold or roll the bag so that there is no air in it. Measure the total mass of the bag, a rubber band, a rubber stopper, and a medicine dropper. Record this mass (the mass of the bag assembly) to the nearest 0.01 g in your data table.

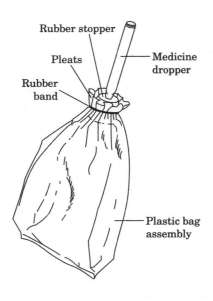

Rubber stopper

Pleats

Medicine
dropper

Rubber
band

Plastic bag
assembly

3. Using tongs, obtain a piece of dry ice about the size of a large pea. Place the dry ice in the bag.

4. As the dry ice sublimes to form gaseous CO_2, gather the open end of the plastic bag in small pleats around the wide end of the rubber stopper. Secure the bag tightly with a rubber band.

5. Handle the bag assembly by the rubber stopper only. Keep the bulb of the medicine dropper off and the dropper pointed up so that any air in the bag will be expelled as the more dense CO_2 fills the bag from the bottom.

6. When all of the dry ice has sublimed and the pressure in the bag is equal to the pressure of the room, the bag will stop expanding. Replace the rubber bulb on the medicine dropper. Measure the mass of the bag and CO_2 to the nearest 0.01 g. Record the mass in your data table.

7. Roll up the bag from the bottom to force out all of the carbon dioxide gas. Following the precautions for hand safety, use a length of rubber tubing to connect the gas jet to the medicine dropper. Open the gas valve slightly to fill the bag with CH_4 from the gas jet. Then turn the gas valve off. In the fume hood, allow the pressure inside the bag to stabilize, and then replace the bulb on the medicine dropper. The bag should be inflated to the same volume as when the CO_2 mass was measured. Obtain the mass of the bag assembly and CH_4 to the nearest 0.01 g and record the mass in your data table.

8. In the fume hood or outdoors, roll up the bag to expel the CH_4.

9. Use a plastic straw to blow air into the bag through the medicine dropper. Allow the pressure of the air to become equal to atmospheric pressure. The bag's volume should be the same as when it was inflated with CO_2 and CH_4. Following the precautions for hand safety, attach a piece of rubber tubing to the dropper and fasten a pinch clamp on the rubber tubing approximately 5 cm from the end of the dropper.

10. Completely fill a 2 L bottle with tap water. Also fill the pneumatic trough with tapwater to the overflow spout. Position the overflow spout over a drain. Place your hand over the mouth of the bottle, invert it, and lower the mouth of the bottle below the surface of the water. Remove your hand from the bottle, and rest the inverted bottle on the shelf. While your lab partner supports the bottle, insert the end of the rubber tubing into the bottle as shown in the illustration on the next page.

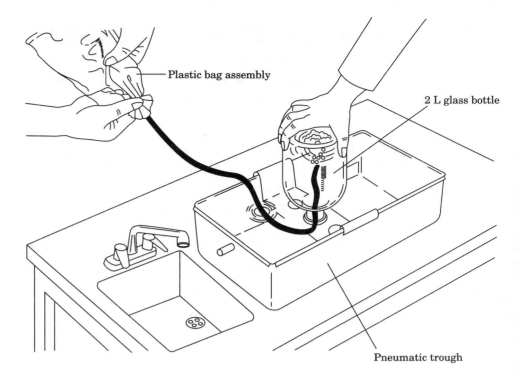

Plastic bag assembly

2 L glass bottle

Pneumatic trough

11. Hold the bag tightly around the rubber stopper to ensure a good seal. Lift the bag above the level of the water and remove the pinch clamp. Very gently squeeze the gas out of the bag. As the bag deflates, crumple or roll it until all of the gas has been removed. Immediately close off the rubber tubing with a pinch clamp and pull the tubing out of the bottle.

12. Place your hand under the water, over the opening of the bottle. Invert the bottle and place it on the lab desk. Use a large graduated cylinder to determine the volume of water required to refill the bottle. Record this volume to the nearest milliliter as the volume of the bag.

13. Record the temperature and pressure in the room.

Cleanup and Disposal

14. Clean all apparatus and your lab station. Make sure that the gas valve at your lab station and the one next to your station are completely shut off before leaving the laboratory. Remember to wash your hands.

Analysis and Interpretation

1. Evaluating Data
Were the masses measured in this lab actual or apparent masses? Explain your answer.

2. Organizing Data
What portion of the mass measured for CO_2 and the assembly was due to CO_2? What portion of the mass measured for CH_4 and the assembly was due to CH_4?

3. Organizing Data
Use the table on the following page to find the density of air at the temperature and pressure of your lab. Calculate the mass of air displaced by the filled bag assembly. The mass of air displaced is equal to the density of air multiplied by the volume of the bag.

Density of Air g/L				
Pressure (mm Hg)	15°C	20°C	25°C	30°C
720	1.16	1.14	1.12	1.10
730	1.18	1.16	1.14	1.12
740	1.19	1.17	1.15	1.13
750	1.21	1.19	1.17	1.16
760	1.23	1.20	1.18	1.16
770	1.24	1.22	1.20	1.18

4. **Organizing Data**
Using your answers to items **1** and **3** and the information given, calculate the actual mass of the CO_2 and CH_4 samples.

5. **Organizing Data**
Determine the ratio of the actual mass of CO_2 to the actual mass of CH_4.

Conclusions

6. **Evaluating Conclusions**
Calculate the molar mass for CO_2, CH_4, and the theoretical mass ratio for CO_2 and CH_4.

7. **Evaluating Methods**
Calculate the percent error for your experimental determination of the mass ratio for CO_2 and CH_4.

8. **Evaluating Conclusions**
Based on your results, is it likely that the tanks, whose contents had a mass ratio of 2.0 : 1.0, contain CO_2 and CH_4? (Hint: consider whether or not your experimental range of error includes the values 2.0 : 1.0 or 1.0 : 2.0.)

9. **Applying Ideas**
Using the periodic table, calculate all possible mass ratios for the following gases: H_2, CH_4, O_2, CO_2. What are the likeliest possible gases contained in the tanks?

Extensions

1. **Evaluating Methods**
Why was it important that you obtain the mass of the methane and carbon dioxide at the same temperature and pressure?

2. **Designing Experiments**
What possible sources of error can you identify in this procedure? If you can think of ways to eliminate them, ask your teacher to approve your suggestions, and run the procedure again.

3. **Applying Ideas**
A student found that the actual mass of a sample of an unknown gas was 1.30 g. A sample of oxygen gas of the same volume at the same temperature and pressure had a mass of 1.48 g. What is the molar mass of the unknown gas?

11-1

Technique Builder

Paper Chromatography

Objectives

Demonstrate proficiency in qualitatively separating mixtures using paper chromatography.

Compare inks by using paper chromatography with a variety of solvents.

Evaluate samples to establish which pen was used on a document.

Situation

Recently, handwriting experts discovered a set of forgeries. Several museums in the United States had been displaying documents supposedly signed by Abraham Lincoln. The FBI suspects that this could be the work of Benny "Fingers" Smithson, who was recently paroled from prison after serving time for his part in a phony Babe Ruth autograph scam. Smithson denies his involvement. A search warrant was issued, and the FBI found three pens in his apartment. The FBI is also investigating another suspect, Thomas Banks, an employee of one of the museums. Three pens from his belongings are also being held as evidence. Before the FBI can press charges, they need conclusive evidence linking the pens and the phony signatures.

Background

Paper chromatography is a method of separating mixtures by using a piece of absorbent paper. In this process, the solution to be separated is placed on a piece of dry filter paper (the stationary phase). A solvent (the moving phase) is allowed to travel across the paper by capillary action. As the solvent is soaked up by the paper, some of the components of the mixture are carried with it. The components of the mixture that are most soluble in the solvent and least attracted to the paper travel the farthest. The resulting pattern of molecules is called a *chromatogram*. In cases where the molecules are easily visible such as in inks, this method distinguishes the components of a mixture.

Problem

To determine which pen was used in the forgery, you must do the following.
- Prepare chromatograms for each of the pens using the two different solvents.
- Prepare chromatograms for different parts of the forged signature using two different solvents.
- Compare the chromatograms and decide which pen is the likeliest match.
- Provide specific examples of similarities between chromatograms, citing measurable points of comparison.

Safety

Always wear goggles and a lab apron to provide protection for your eyes and clothing. If you get a chemical in your eyes, immediately flush it out at the eyewash station while calling to your teacher. Know the locations of the emergency lab shower and eyewash and how to use them.

Do not touch any chemicals. If you get a chemical on your skin or clothing, wash it off at the sink while calling to your teacher. Make sure you carefully read the labels and follow the directions on all containers of chemicals that you use. Never return leftovers to their original containers; take only small amounts to avoid wasting supplies.

Because the isopropanol is flammable, no Bunsen burners, hot plates, or other sources of heat should be in use in the room during this lab. Carry out all work with isopropanol in a hood.

Always clean up the lab and all equipment after use, and dispose of substances according to proper disposal methods. Wash your hands thoroughly before you leave the lab after all lab work is finished.

Materials

- Distilled water
- Isopropanol
- Filter papers, 12 cm, 4
- Filter paper wicks, 2 cm equilateral triangles, 2
- Forged signature samples
- Paper clips, 2
- Pens, black ink, 6
- Pencil
- Petri dish with lid
- Ruler

Preparation

1. *Organizing Data*
   Prepare your data table with seven columns and eight rows. Label the boxes in the top row *Pen number, H₂O—center, H₂O—middle, H₂O—edge, Isopropanol—center, Isopropanol—middle,* and *Isopropanol—edge*. In the second through eighth columns of the first column, add the labels *Smithson 1, Smithson 2, Smithson 3, Banks 4, Banks 5, Banks 6,* and *Forgery*.

Technique

2. Use a pencil to sketch a circle about the size of a quarter in the center of the piece of filter paper. Write the numbers 1–6 in pencil around the inside of this circle, as shown in the illustration on the next page.

3. On the circle beside the number 1, use pen number 1 to make a large dot. Use pen number 2 to make a dot beside number 2, and repeat this procedure for each pen.

4. Repeat steps **2** and **3** with a second piece of filter paper. One will be used with water as a solvent, and the second will be used with isopropanol as a solvent.

5. Roll up the triangle of filter paper to be used as a wick. Use the pencil to poke a small hole in the center of the first marked piece of filter paper. Insert a rolled-up piece of the wick through the hole, as shown in the illustration on the next page.

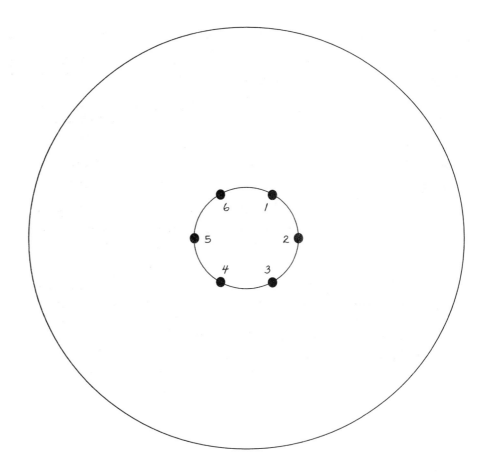

6. Fill the petri dish to the halfway point with water. Set the wick of the filter paper into this water, as shown in the illustration, and wait for the chromatogram to develop.

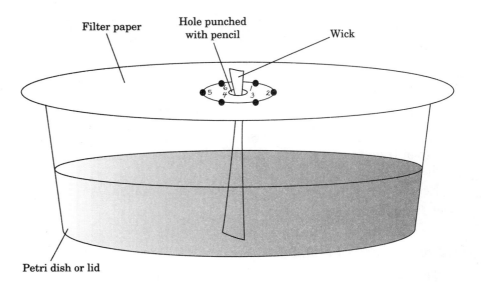

7. Repeat steps **5** and **6** with the second piece of filter paper. Instead of water, use isopropanol to fill the petri dish lid to the halfway point.

8. Allow the chromatograms to develop for approximately 15 min, or until the solvent is about 1 cm away from the outside edge of the paper. Remove each piece of filter paper from the petri dish and lid, and allow them to dry.

9. Record the colors that have separated on the chromatogram from each of the six different black inks in your data table. You may either describe the colors or use colored pencils to record this information.

10. Take the piece of the forged signature, and choose two segments that can be used to make chromatograms. (Hint: consider in which direction the ink will travel after the wick brings solvent onto the paper.) Cut out the two segments, leaving as much blank paper attached to each one as possible.

11. With one piece of the paper, make a chromatogram using the water in the petri dish. Do not allow the ink from the paper to come into direct contact with the solvent. Instead, use a wick as before. Unbend the paper clip, and use it to support the strip with the signature, as shown in the illustration. Then make a chromatogram using the isopropanol in the petri dish lid with the other piece of the paper.

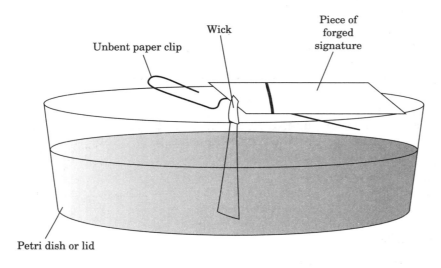

Unbent paper clip

Wick

Piece of forged signature

Petri dish or lid

12. After 15 min or when the solvent is about 1 cm away from the top of the paper, remove each chromatogram from the petri dish or lid, and allow them to dry.

13. Record the colors from the pieces of the signature and their arrangement in your data table.

Cleanup and Disposal

14. The water may be poured down the sink. Chromatograms and other pieces of filter paper may be discarded in the trash can. The isopropanol solution should be placed in the waste disposal container designated by your teacher. Clean up your equipment and lab station. Thoroughly wash your hands after completing the lab session and cleanup.

Analysis and Interpretation

1. **Organizing Ideas**
   Draw Lewis dot structures for water, H_2O, and isopropanol, $CH_3CHOHCH_3$.

2. **Analyzing Ideas**
   Analyze the bonding and structure of water and isopropanol molecules, as discussed in Chapter 11. What types of substances are most likely to dissolve in each one? (Hint: be sure to refer to page 426 when working with isopropanol.)

3. **Applying Ideas**
   Some components of ink are minimally attracted to the stationary phase and very soluble in the solvent. Where on the filter paper should these components be located in the final chromatogram?

4. **Relating Ideas**
   The diffusion of a solute in a solvent is somewhat similar to the diffusion of gases discussed in Chapter 10. Where on the filter paper do you expect the larger molecules to be located in the final chromatogram?

5. **Evaluating Methods**
   Explain why the labels numbering the pen spots were written in pencil. (Hint: recall what you know from Chapter 6 about the nature of the bonds and structure in the graphite that is found in pencils.)

6. **Predicting Outcomes**
   What would have happened to the chromatogram if the process had not been stopped after 15 min, but instead was allowed to proceed overnight?

7. **Predicting Outcomes**
   At Bryan High School, Armando was performing the chromatogram experiment. He thought it was going too slow and decided to work without a wick. He dipped the sample into the solvent so that the ink dot was even with the top of the water. But when he finished, the chromatogram was too faint to be seen. Explain why Armando's experiment failed.

Conclusions

8. **Analyzing Conclusions**
   Are the properties of the component that traveled the furthest in the water chromatogram likely to be similar to the properties of the component that traveled the furthest in the isopropanol chromatogram? Explain your reasoning.

9. **Analyzing Conclusions**
   Which pen was used for the forged signature? Explain your reasoning, giving specific examples and providing quantitative data as possible (Hint: measure the distances that each component traveled to gather quantitative data.)

Extensions

1. **Designing Experiments**
   How would you improve the efficiency of the separation? If you can think of a way to make the technique work better, ask your teacher to approve your plan, and run the procedure again.

2. Predicting Outcomes

Amino acids are organic chemicals that are monomers of proteins. They each have at least one amine group and one carboxylic acid group, each of which can carry charge in certain solutions. A scientist is analyzing a sample that contains a mixture of the four amino acids shown below. She can detect the presence of any amino acid by spraying it with an indicator called ninhydrin. She makes two chromatograms, one with water and another with isopropanol, using different portions of the sample. In what order would the amino acids be after each chromatogram was dried and sprayed with ninhydrin? (Hint: it may be helpful to refer to your answers to items **4** and **8**. Also consider what you know about solubility rules from Chapter 11.)

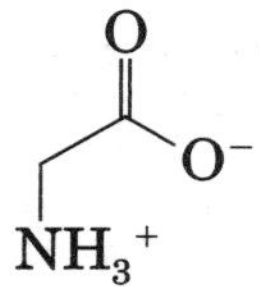

Glycine

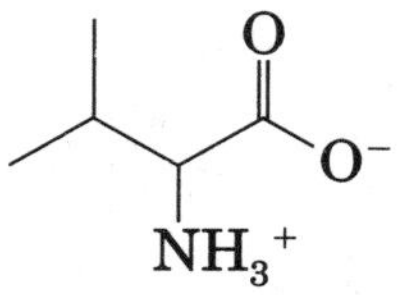

Valine

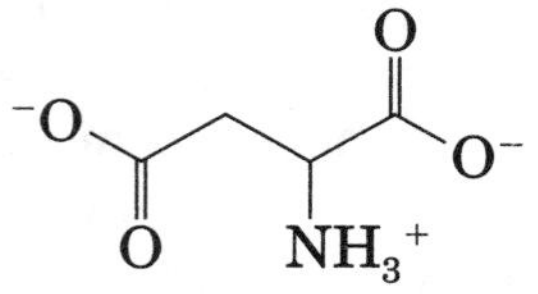

Aspartic acid

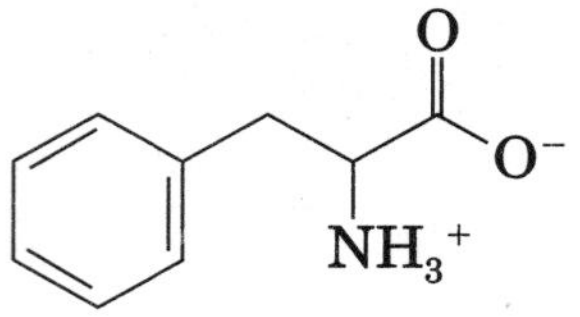

Phenylalanine

3. Applying Ideas

In a gel chromatography apparatus, a polymeric gel that has many pores, or openings, is packed into a long tube or column. A solvent is continually forced under low pressure through the tube. The gel serves as the stationary phase, and the solvent is the mobile phase. A sample of a mixture can be injected into the solvent just before it enters the column, and it will be separated into its component parts. Each component of the sample will take a different amount of time to pass through the column. If the solvent used is benzene, in what order will the plant pigments shown below pass through the column? (Hint: it may be helpful to refer to your answers to items **4** and **8**. Also consider what you know about solubility rules from Chapter 11.)

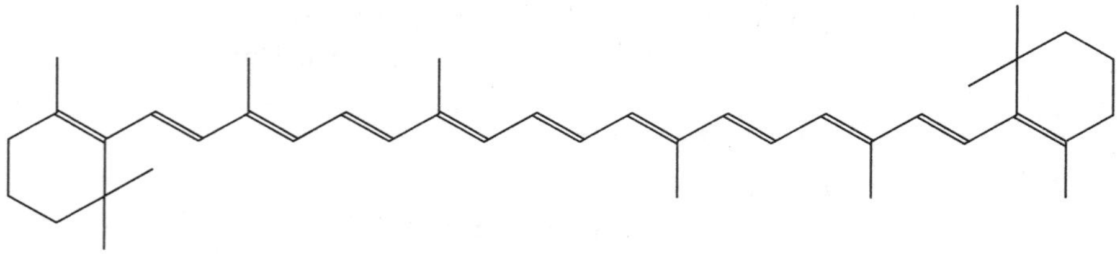

Beta carotene (yellow/orange)

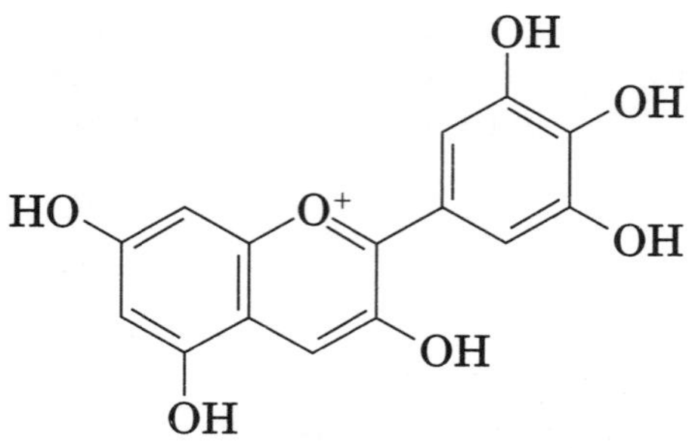

Delphinidin (blue/violet)

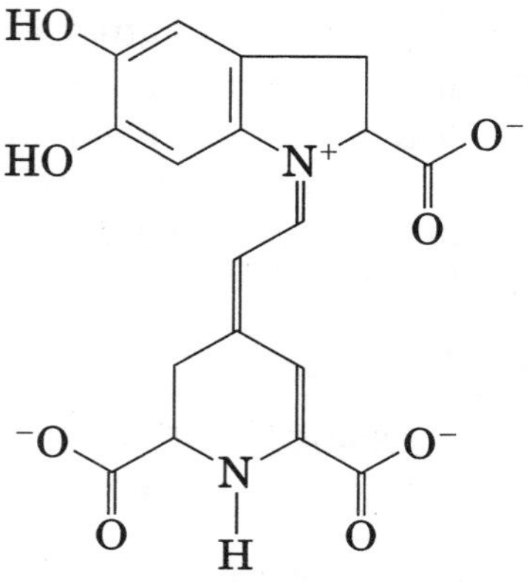

Betanidin (reddish)

INVESTIGATION 11-1

Problem
Solving

Paper Chromatography— Forensic Investigation

March 23, 1995

Marissa Bellinghausen
Director of Investigations
CheMystery Labs, Inc.
52 Fulton Street
Springfield, VA 22150

Dear Ms. Bellinghausen:

Late last night, someone broke into a display case at Frederick Douglass High School and stole a rare manuscript of a speech in Douglass's own writing that was part of a special exhibit.

While our detectives were examining the break-in area this morning, a hand-lettered ransom note asking for $100 000 was placed in the principal's mailbox.

We have been unable to pin down the time of the theft, but the ransom note was placed in the mailbox between 10:14, when the principal emptied it, and 10:25, when he noticed something else there.

Only four people had entered the hallway where mailbox is. Each of them agreed to submit to a search, but the missing manuscript was not found. However, each of the possible suspects was carrying a black pen that could have been used on the ransom note.

If we can obtain evidence proving which suspect wrote the note, we believe the perpetrator will surrender the priceless manuscript before it is damaged or sold.

Sincerely,

Alyssa Hartman

Alyssa Hartman
Captain, Eleventh Precinct
Springfield Police Department

Memorandum

Date: March 24, 1995
To: Janey Markowitz
From: Marissa Bellinghausen

CheMystery Labs, Inc.
52 Fulton Street
Springfield, VA 22150

We need to act <u>quickly</u>. If we can demonstrate that we can provide <u>accurate</u> and <u>prompt results</u>, more work from law enforcement could come our way. Due to budget cutbacks, the police department can afford to spend only <u>$75 000</u> on this job. Because of the number of tests that must be performed on the evidence, you will be given only a very <u>small piece of the</u> <u>sample</u>, <u>so plan carefully</u>!

Before you begin the job, Captain Hartman says the district attorney would like to make sure our <u>technique will hold up in court</u>. Prepare the following for me to send to them.

- <u>detailed one-page plan</u> for the procedure and all <u>necessary data tables</u>
- <u>detailed list</u> of the equipment and materials you will need, along with the <u>individual and total</u> <u>costs</u>

After I tell you that the district attorney is satisfied, you may begin your lab work. When your research team has completed the lab work, you should prepare a report in the form of a <u>two-page letter</u> to Captain Hartman. Remember that this <u>document</u> will become part of the <u>police report</u> and <u>also could become evidence</u> in a later trial. It should include all of the following.

- <u>which pen</u> was used to write the ransom note
- <u>summary paragraph</u> of how you solved this problem
- discussion of the <u>certainty</u> of your method
- detailed and organized <u>analysis</u> of the sample
- detailed and organized <u>data table</u>
- detailed <u>invoice</u> including costs for equipment and supplies as well as services rendered

Required Precautions

Always wear goggles and a lab apron to provide protection for your eyes and clothing.

Do not touch any chemicals. If you get a chemical on your skin or clothing, wash it off at the sink while calling to your teacher.

Isopropanol is flammable. No burners, flames, hot plates, or other heat sources should be present in the room while isopropanol is being used. Keep isopropanol in the hood.

Always wash your hands thoroughly when finished.

Spill Procedures/ Waste Disposal Methods

- In case of a spill, follow your teacher's instructions.
- Put excess isopropanol in the container designated by your teacher. Do not pour it down the sink.
- Any of the water may be poured down the sink.
- The chromatograms and filter paper wicks may be discarded in the trash can.

References

Carlo Avena, the chemistry teacher, came in at 10:17 along with Johnny Pudstrom to discuss with the principal Johnny's latest excuse for not having his lab goggles, homework, or books. Mr. Avena, who owns Pen A, is said to be jealous that the biology teacher was named science department chair. Johnny has been in a number of scrapes this year, and the principal has put him on probation. Johnny owns Pen B, which was the only school supply he had with him.

Laney Femberton, a senior, brought in the collected attendance reports for the third period at 10:20. Laney is eager to find a way to make enough money so that she can quit school and her after-school job and devote more time to the beauty contest circuit. Laney owns Pen C.

The last possible suspect is the principal, Joshua Skinner. The superintendent has announced an audit of the school's finances in order to put to rest once and for all rumors of mismanagement. Skinner owns Pen D.

MATERIALS for Forensic Investigation

REQUIRED ITEMS

(You must include all of these in your budget.)

Lab space/fume hood/utilities	15 000/day
Standard disposal fee	20 000/L of solution

REAGENTS AND ADDITIONAL EQUIPMENT

(Include in your budget only what you will need.)

Isopropanol	1 000/mL
150 mL beaker	1 000
250 mL beaker	1 000
400 mL beaker	2 000
250 mL Erlenmeyer flask	1 000
100 mL graduated cylinder	1 000
Aluminum foil	$1\ 000/cm^2$
Balance	5 000
Cotton swab	500
Desiccator	3 000
Filter paper	500/piece
Paper clips	500/box
Petri dish and lid	1 000
Pipet or medicine dropper	1 000
Ruler	500
Stopwatch	5 000
Wax pencil	500

* *No refunds on returned chemicals or unused equipment.*

FINES

OSHA safety violation	2 000/incident

11-2 Testing for Dissolved Oxygen

Technique Builder

Situation

The company you work for has been hired as an expert witness in a lawsuit. The local chapter of Bass Anglers Unlimited has been disturbed by recent declines in the population of bass in Pulaski Lake. There have been several fish kills, in which large numbers of fish die at the same time and float to the surface of the lake, creating a terrible smell and fouling the water. The fishermen claim that the fish kills began shortly after the R. C. Throckmorton Power Plant came on line, and they are seeking a court order to shut down the plant. The machinery in the power plant uses water from the lake as a coolant and then returns the water to the lake.

The fishermen say that something in the returned water is killing the fish. The utility company operating the power plant points out that they use a closed system that prevents the water from coming into direct contact with the machinery in the plant. They say that the water is just as pure when it comes out; it's just a little warmer. The court has asked you to investigate whether there is a scientific basis for the fishermen's claim.

Background

Fish rely on the oxygen dissolved in water to live. The water passes through their gills, which remove the oxygen. The less oxygen in the water, the harder the gills have to work to get enough oxygen to keep the fish alive. The normal lake temperature is between 15°C and 17°C. Another witness, a biology professor specializing in ichthyology (the study of fish) has testified already that if the oxygen content of the water dips below 90% of the value at these temperatures, the fish will suffer long-term damage or death. The temperature of the water returned to the lake from the power plant is 28°C. To measure the change in dissolved oxygen content over this temperature range, you can use tap water, which has a mineral content similar to lake water.

Problem

In order to evaluate these claims, you will need to do the following.
- Prepare water samples at several different temperatures.
- Measure the dissolved oxygen content in ppm for each one.
- Graph the relationship between the solubility of dissolved oxygen and temperature.
- Extrapolate to determine the solubility of dissolved oxygen at 16°C and 28°C.
- Compare the solubilities to determine if there is enough difference to cause damage to the fish.

Objectives

Measure the concentration of oxygen in a sample of water.

Graph the relationship between the concentration of a gas and temperature.

Infer a general rule of thumb for gas solubilities and temperature.

Relate changes in gas solubility to a fish kill.

Materials

- Ice
- 600 mL beaker
- Beaker tongs
- Hot mitt
- Jars with screw-on lids, 4

Bunsen burner option

- Bunsen burner and related equipment
- Ring stand
- Ring clamp
- Wire gauze with ceramic center

Hot plate option

- Hot plate

LEAP System option

- LEAP System
- LEAP dissolved oxygen probe
- LEAP thermistor probe

Alternative option

- Dissolved oxygen test kit or dissolved oxygen meter
- Thermometer, non-mercury

Safety

 Always wear goggles and a lab apron to provide protection for your eyes and clothing. If you get a chemical in your eyes, immediately flush it out at the eyewash station while calling to your teacher. Know the locations of the emergency lab shower and eyewash and how to use them.

 When you use a Bunsen burner, confine any long hair and loose clothing. Do not heat glassware that is broken, chipped, or cracked. Use tongs or a hot mitt to handle heated glassware and other equipment because hot glassware does not always look hot. If your clothing catches on fire, WALK to the emergency lab shower, and use it to put out the fire.

 Always clean up the lab and all equipment after use, and dispose of substances according to proper disposal methods. Wash your hands thoroughly before you leave the lab after all lab work is finished.

Preparation

1. **Organizing Data**
   Prepare a data table in your lab notebook with two columns and five rows. In the first row, label the two columns *Temp. (°C)* and *Dissolved O₂ (ppm)*.
2. Label four jars *Ice water*, *Room temp.*, *50°C*, and *100°C*.

Technique

Sample Preparation

3. Add approximately 50 g of ice to 100 mL of tap water in the *Ice water* jar. Let the ice-water mixture stand for 5 min. Measure the temperature with a thermometer or LEAP thermistor probe to the nearest 0.1°C. The temperature should be near 4°C. Record the temperature in your data table. Add more ice until the jar is filled to the rim. Screw on the top and place the sample in an ice chest or refrigerator overnight.

4. Fill the *Room temp.* jar with tap water. Do not seal the jar. Let it sit out open overnight where it will not be disturbed. You will measure the temperature of this sample tomorrow.

5. If you are using a Bunsen burner, set up the ring stand, ring, and wire gauze over the burner so that they will hold a beaker. If you are using a hot plate, continue with step **6.**

6. Pour approximately 450 mL of tap water into a 600 mL beaker, and gently heat to about 50°C. Maintain 50°C as closely as possible for 5 min. Measure the temperature of the water with a thermometer or LEAP thermistor probe to the nearest 0.1°C. Record the temperature in your data table. Using beaker tongs to hold the hot beaker, gently fill the jar labeled *50°C* with the warm water. Fill the container to the rim. Screw on the top, and store in a safe place.

7. Heat the remaining water to boiling (approximately 100°C). Allow the water to boil for approximately 30 min. Measure the temperature of the water with a thermometer or LEAP thermistor probe to the nearest 0.1°C. Record the temperature in your data table. Using beaker tongs to hold the hot beaker, gently fill the jar labeled *100°C* with the boiling water. Fill the container to the rim. Using a hot mitt, screw on the top and store in a safe place.

Sample Testing

8. On the following day, retrieve your samples.

9. Measure the temperature of the water in the *Room temp.* jar with a thermometer or LEAP thermistor probe to the nearest 0.1°C. Be sure to disturb the water as little as possible. Record the temperature in your data table.

10. If you are using a dissolved oxygen probe or meter, it may be necessary to calibrate it using a standardized solution. Ask your teacher for instructions. If your teacher indicates that this is unnecessary, continue with step **11.**

11. Disturbing the water as little as possible, measure the dissolved oxygen content of each sample, and record the value to the nearest 0.1 ppm in your data table. If you are using a dissolved oxygen test kit, you will be comparing the colors of standard solutions to the colors of your tested solutions. Estimate the measured concentrations to the nearest 0.5 ppm.

Disposal

12. If you used chemical test kits, dispose of the ampuls and reagents in the containers designated by your teacher. Rinse the samples down the drain. Clean all equipment.

Analysis and Interpretation

1. **Organizing Data**
   Make a graph of your data, with temperature (in °C) plotted on the x-axis and O_2 concentration (in ppm) plotted on the y-axis. Draw a straight line that best fits the data.

2. **Interpreting Graphics**
   Refer to the graph from item **1,** and explain in your own words the relationship between temperature and the solubility of oxygen in water.

3. **Analyzing Methods**
   Why was it important to disturb the water as little as was possible when measuring the dissolved oxygen content?

4. **Applying Models**
   Often, when water is heated, small bubbles appear long before boiling begins. Explain why these bubbles form. (Hint: use the concept of solubility in your explanation.)

Conclusions

5. **Interpreting Graphics**
   Using the graph from item **1,** extrapolate to determine the concentration of O_2 in the water at 16°C and 28°C.

6. **Interpreting Graphics**
   Using the format shown below, determine the equation for the line you drew on the graph in item **1**. (Hint: if you have a graphics calculator, use the $\boxed{\text{STAT}}$ mode to enter your data and make a linear regression equation using the LinReg function from the STAT menu.)

   $$O_2 \text{ conc.} = m(\text{temp.}) + b$$

7. **Analyzing Conclusions**
   Calculate the expected concentration of O_2 in the water at 16°C and 28°C, using the equation from item **6**.

8. **Evaluating Conclusions**
   Using the values from item **5** as the accepted values, calculate the percent error for the values calculated in item **7**.

9. **Evaluating Conclusions**
   Earlier in the trial, the ichthyologist testified that fish receiving less than 90% of the oxygen available at 16°C would suffer long-term damage. From item **5** and item **7**, determine what percentage of the oxygen available at 16°C is available at 28°C. Is it likely that the water is damaging to the fish?

Extensions

1. **Evaluating Conclusions**
   During cross-examination, the attorney for the power plant suggests that because you tested tap water instead of water taken directly from the power plant, your solubility results are irrelevant, and the power plant's water could have even more oxygen in it than the original cold water. Is the attorney correct? Why or why not?

2. **Evaluating Conclusions**
   Later in the cross-examination, the attorney for the power plant asks whether your results establish conclusively that the fish that were killed died because of the warmer water and for no other reasons. What do you say?

3. **Designing Experiments**
   Assuming that you had unlimited laboratory resources and access to all of the lake, what other tests would you perform to be more certain of your results?

4. **Evaluating Methods**
   The judge has asked for the opinions of all expert witnesses about a proposed settlement. The power plant proposes to insert a device to bubble oxygen through the warm water as it is released into the lake. Will this solve the problem? Explain why or why not, using the principles of solubility.

5. **Relating Ideas**
   The power plant supplies 10 000 households with their electricity. The judge has asked for possible solutions that will keep the power plant working and prevent further damage to the fish. What do you suggest?

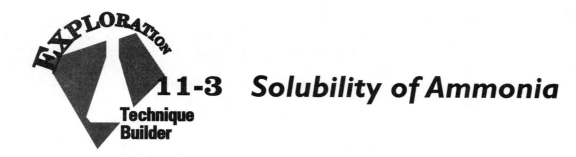

11-3 Solubility of Ammonia

Technique Builder

Situation

Sparkle-Clean Industries has contacted you about one of their products, a household window cleaner with ammonia, NH_3. Sparkle-Clean did not have a product for this market until they bought out a competitor. The competitor purchased gaseous NH_3 from a local gas products supplier. Then the NH_3 was dissolved in water at room temperature as the first and most difficult step in the manufacturing process.

Sparkle-Clean's manufacturing equipment has several features that the competitor did not have. One piece of equipment can maintain liquids at temperatures from 0°C to 80°C. Perhaps changing the temperature could improve the competitor's procedure so that more NH_3 would dissolve in the water. However, the equipment would need expensive adjustments to keep the hazardous NH_3 gas under control. To determine whether the savings will justify the initial expense, they want you to model the process on a small scale, determine the best temperature, and estimate how much NH_3 dissolves at that temperature.

Background

From Chapter 11, you know that most gases dissolve better in water at cooler temperatures, provided that the pressure remains the same. This is the opposite of the pattern observed for most solids. In the text, the example of a soda bottle was given. The warmer the soda, the flatter it will taste because the dissolved CO_2 that provides the "fizz" will bubble out of solution. You must test the NH_3 to see if it follows the same pattern. Measuring how long it takes for a given amount of NH_3 to dissolve can also provide a rough measure of how much more efficient the process will be at a different temperature.

Problem

To determine the optimum temperature for this procedure, you must do the following.
- Prepare some NH_3 gas by heating a solution of aqueous NH_3.
- Test the NH_3 by dissolving it in water at various temperatures.
- Compare efficiencies of the dissolving process for the NH_3 in the different water samples by measuring the time for it to dissolve completely.

Objectives

Prepare a small amount of ammonia gas in a container.

Calculate amount of ammonia present, given volume, temperature, and pressure.

Test how well the ammonia dissolves in water at different temperatures.

Relate changes in volume of a gas to the removal of gas particles as they dissolve in a solution.

Materials

- NH$_3$(aq), concentrated
- Distilled water
- Ice
- 250 mL beakers, 6
- 400 mL beaker
- Pipet, thin-stem
- Pipet, wide-stem
- Ruler, metric
- Scissors
- Stopwatch or clock with second hand
- Thermometer, non-mercury

Bunsen burner option

- Bunsen burner and related equipment
- Ring stand
- Ring clamp
- Wire gauze with ceramic center

Hot plate option

- Hot plate

LEAP System option

- LEAP System
- LEAP thermistor probe

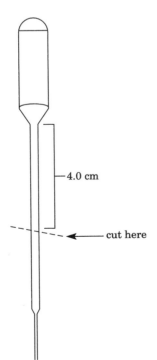

4.0 cm

cut here

Safety

Always wear goggles and a lab apron to provide protection for your eyes and clothing. If you get a chemical in your eyes, immediately flush it out at the eyewash station while calling to your teacher. Know the locations of the emergency lab shower and eyewash and how to use them.

Because the procedure involves poisonous ammonia fumes, the procedure must take place in a properly functioning fume hood. Make sure that the fume hood is turned on before starting the reaction.

Do not touch any chemicals. If you get a chemical on your skin or clothing, wash it off at the sink while calling to your teacher. Make sure you carefully read the labels and follow the directions on all containers of chemicals that you use. Never return leftovers to their original containers; take only small amounts to avoid wasting supplies.

Ammonia is a base. Call your teacher in the event of a spill of the NH$_3$ solution. Acid or base spills should be cleaned up promptly, according to your teacher's instructions.

When you use a Bunsen burner, confine any long hair and loose clothing. Do not heat glassware that is broken, chipped, or cracked. Use tongs or a hot mitt to handle heated glassware or other equipment that has been heated because hot equipment does not always look hot. If your clothing catches on fire, WALK to the emergency lab shower and use it to put out the fire.

Always clean up the lab and all equipment after use, and dispose of substances according to proper disposal methods. Wash your hands thoroughly before you leave the lab after all lab work is finished.

Preparation

1. **Organizing Data**
   Prepare a data table in your lab notebook with five rows and two columns. Label the columns in the first row *Temp. (°C)* and *Time (s)*. You will also need room beneath the data table to record the total volume of the ammonia sample.

2. Cut off the tip of the wide-stem pipet at an angle so that the stem is as close to 4 cm long as possible. This will be the collecting pipet in which the ammonia gas is collected.

3. If you were given the correct type of bulb and followed the directions in step **2** exactly, the volume of the bulb should be about 5.0 mL. If you used different equipment, ask your teacher how to measure the volume of the bulb. You will need a 10 mL or 25 mL graduated cylinder to measure the bulb's volume.

4. One of the 250 mL beakers will already be labeled *Conc. NH₃(aq)* by your teacher. It should already contain concentrated ammonia solution. Label the remaining five beakers *Holder, 80°C, 50°C, Room temp.,* and *0°C*.

Technique

5. While working in the hood, prepare a warm-water bath with about 350 mL of water in the 400 mL beaker using a Bunsen burner and a ring stand or a hot plate. The temperature of the water should be about 80°C.

6. Pour 200 mL of the hot water into the *80°C* beaker. Keep it under the hood and not far from the warm-water bath. Refill the warm-water bath.

7. Working in the hood, fill the narrow-stem pipet about halfway with concentrated aqueous ammonia from the *Conc. NH₃(aq)* beaker. This pipet is the gas generating pipet. Turn it upside down (bulb end down). Keep it under the hood at all times.

8. Loosely cover the tip of the generating pipet with the bulb of the collecting pipet, as shown in the diagram. Be sure that the air in the collecting pipet can escape as the NH_3 enters and displaces the air.

9. Hold the pipet assembly with the bulb of the generating pipet dipped into the warm-water bath under the hood so that the aqueous NH_3 warms. As the solution warms, NH_3 gas is driven off into the collecting bulb. You will notice the aqueous NH_3 solution bubbling. Squeeze the collecting pipet several times to expel all of the air and allow the NH_3 gas to completely fill the collecting pipet.

10. Measure the temperature of the water in the *80°C* beaker with a thermometer or a LEAP System thermistor probe to the closest 0.1°C. Record the temperature in your data table.

11. After 2–3 min, remove the generating pipet and collecting pipet from the warm-water bath. Place the generating pipet with the bulb end down into the *Holder* beaker.

12. Using a stopwatch or having your partner keep an eye on a clock with a second hand, place the tip of the collecting pipet into the water in the *80°C* beaker. Record your observations in the data table. Measure and record how many seconds it takes until the gas is completely dissolved and the bulb is filled with water.

13. Pour about 200 mL of the water from the warm-water bath into the *50°C* beaker. Refill the warm-water bath.

14. Using the procedure from steps **7–9**, work under the hood to collect another bulb of NH_3.

15. Add ice to the water until the temperature of the water in the *50°C* beaker is close to 50°C. Empty some of the water until about 200 mL remains.

16. Measure the temperature of the water in the *50°C* beaker with a LEAP System thermistor probe or a thermometer, and record it in the data table. Return this beaker to the hood.

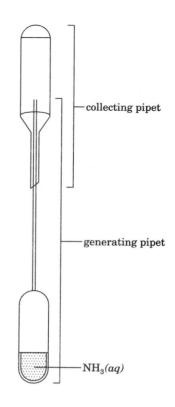

collecting pipet

generating pipet

$NH_3(aq)$

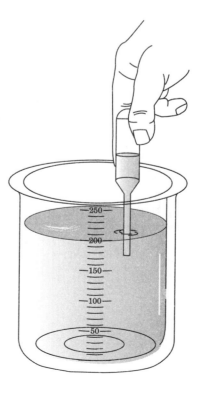

HRW material copyrighted under notice appearing earlier in this work.

17. Using a stopwatch or having your partner keep an eye on a clock with a second hand, place the tip of the collecting pipet into the water in the *50°C* beaker. Record your observations in the data table. Measure and record how many seconds it takes until the gas is completely dissolved.

18. Repeat steps **14–17**, working under the hood with 200 mL of water at room temperature in the *Room temp.* beaker and with 200 mL of an ice-water mixture in the *0°C* beaker.

Disposal

19. The solutions of aqueous ammonia should be left in a disposal container in the hood. Clean all other equipment after use.

Analysis and Interpretation

1. *Analyzing Methods*
   How can you be certain that the air will be displaced out of the bulb instead of the NH_3? (Hint: air is mostly O_2 and N_2. Of the three gases, which two are likely to be the densest?)

2. *Analyzing Methods*
   Use your own words to explain how and why the NH_3 gas was generated when the NH_3 solution was heated.

3. *Analyzing Data*
   Using the ideal gas law, calculate the number of moles of NH_3 present from the initial volume in the bulb and pipet tip before it is tested in the water. (Hint: you measured the temperature of room temperature water, and the pressure is about 1.0 atm in the lab.)

4. *Analyzing Data*
   Using the periodic table and your answer from item **3**, calculate the mass of NH_3 present in the bulb at the beginning.

5. *Analyzing Information*
   Explain why the water was pushed into the collecting bulb. (Hint: what happened to some of the NH_3 when it came into contact with water?)

Conclusions

6. *Evaluating Conclusions*
   Which temperature would be the most effective for Sparkle-Clean to use in their new process? Explain why.

7. *Evaluating Conclusions*
   If the process at room temperature dissolves 500. g of ammonia gas in 15 min, approximate how long it would take to dissolve 500. g of ammonia gas at the temperature from item **6.**

8. *Evaluating Conclusions*
   Can you tell from your data how much more NH_3 will dissolve at the most effective temperature?

9. *Applying Ideas*
   Using information on solubility and polarity from Chapter 11, explain why this procedure works for dissolving NH_3 in water, but not for dissolving CO_2 to make a soft drink.

Extensions

1. Applying Ideas
Actually, the NH_3 that was collected was not pure. Because it was collected over water, some of the gas in the bulb was water vapor. Using the vapor pressure value for water that matches the conditions in your lab, recalculate the amount of NH_3 in the bulb before each of the trials. (Hint: 1.0 atm = 101.325 kPa)

Temperature (°C)	Partial Pressure of H_2O (kPa)
16	1.82
17	1.94
18	2.06
19	2.19
20	2.34
21	2.49
22	2.64
23	2.81
24	2.98
25	3.17
26	3.36
27	3.56
28	3.78

2. Predicting Outcome
According to the discussion of Henry's law in Chapter 11, the solubility of a gas is directly proportional to the partial pressure of the gas over the surface of the water. Sparkle-Clean can run the procedure at pressures of 1.0, 1.5, and 2.0 atm. Which should they pick to maximize the amount of NH_3 dissolving in the water? How much NH_3 should dissolve at this pressure?

3. Designing Experiments
What possible sources of error can you identify with this procedure? If you can think of ways to eliminate them, ask your teacher to approve your plan, and run the procedure again.

4. Research and Writing
Studies indicate that much of the CO_2 that is generated by living things and by the combustion of fossil fuels is dissolving in ocean water. Assuming this is true, explain how the CO_2 levels in the atmosphere would be affected if the environmental models that predict a period of global warming are correct.

12-1 Solubility Product Constant

Technique Builder

Objectives

Prepare a saturated solution of sodium chloride.

Determine the mass of sodium chloride dissolved in a volume of saturated solution.

Calculate the solubility of sodium chloride.

Determine the solubility product constant of sodium chloride.

Demonstrate the effect of the addition of a common ion to a saturated solution.

Situation

Juliette Brand Foods is planning to add dill pickles to their product line. Pickled foods require soaking or packing in a concentrated solution of sodium chloride called brine. The company intends to make their own brine in large quantities. To help them set up their process, they have asked your company to determine the mass of NaCl that should be added to water in a 4000.0 L vat at room temperature so that the solution will have the highest possible concentration with no NaCl remaining undissolved. Juliette Brand Foods is also considering the use of sodium citrate, $Na_3C_6H_5O_7$, as a flavoring agent. They need you to determine what effects, if any, on the solubility of NaCl the addition of $Na_3C_6H_5O_7$ will have.

Background

As it was defined in Chapter 11, a saturated solution contains the maximum amount of solute that can dissolve when added to a solvent at a given temperature. As discussed in Chapter 12, saturated solutions usually contain an excess of undissolved solid so that the ions in solution are at equilibrium with the undissolved solid. The equilibrium constant that describes this situation is called the solubility product constant, K_{sp}. For salts composed of ions with single positive or negative charges, the K_{sp} is the product of the concentrations of undissolved ions. If either of the ion concentrations is increased after a saturated solution is formed, the value of the K_{sp} will be exceeded, and precipitation will occur until equilibrium is restored. This phenomenon is called the "common-ion effect."

Problem

To answer your client's questions you need to do the following.
- Make a saturated solution of NaCl at room temperature.
- Calculate the concentration in moles per liter of each ion in the solution.
- Write the equilibrium expression and calculate the K_{sp} of NaCl at room temperature.
- Observe the effect of adding $Na_3C_6H_5O_7$ to a saturated solution of NaCl.
- Use your data to calculate the mass of NaCl necessary to fill the pickling vat with 4000.0 L of saturated solution.

Safety

Always wear goggles and a lab apron to provide protection for your eyes and clothing. If you get a chemical in your eyes, immediately flush it out at the eyewash station while calling to your teacher. Know the locations of the emergency lab shower and eyewash and how to use them.

Do not touch any chemicals. If you get a chemical on your skin or clothing, wash it off at the sink while calling to your teacher. Make sure you carefully read the labels and follow the directions on all containers of chemicals that you use. Do not taste any chemicals or items used in the laboratory. Never return leftovers to their original containers; take only small amounts to avoid wasting supplies.

When you use a Bunsen burner, confine any long hair and loose clothing. Do not heat glassware that is broken, chipped, or cracked. Use tongs or a hot mitt to handle heated glassware or other equipment that has been heated because hot equipment does not always look hot. If your clothing catches on fire, WALK to the emergency lab shower, and use it to put out the fire.

Always clean up the lab and all equipment after use, and dispose of substances according to proper disposal methods. Wash your hands thoroughly before you leave the lab after all lab work is finished.

Materials

- Distilled water
- NaCl, 15 g
- $Na_3C_6H_5O_7 \cdot 2H_2O$, 0.50 g
- 150 mL beakers, 2
- 10 mL graduated cylinder
- 25 mm × 100 mm test tube
- Balance
- Crucible tongs
- Evaporating dish
- Glass stirring rod
- Spatula
- Weighing paper

Bunsen Burner option

- Bunsen burner and related equipment
- Ring stand and ring
- Wire gauze with ceramic center, 2

Hot Plate option

- Hot plate

Preparation

1. **Organizing Data**
   Prepare a table for data and observations. You will need space to record the mass of the empty evaporating dish, the volume of saturated NaCl solution, and the mass of the evaporating dish and NaCl. You will also need a space to record the mass of $Na_3C_6H_5O_7 \cdot 2H_2O$ used. Be certain there is additional room for recording your observations from step **11.**

2. Make sure that your equipment and tongs are very clean for this work so that you will get the best possible results. Remember that you will need to cool the evaporating dish before measuring its mass. ***Never put a hot evaporating dish on a balance; it may damage the balance, and the currents of hot air will cause an error in the mass measurement.***

Technique

3. Prepare a saturated solution of NaCl by adding 15 g of the salt to 25 mL of water in a beaker. Stir constantly until it appears that no more salt will dissolve. There should be some solid on the bottom of the beaker. Decant the saturated solution into another beaker, making certain that no solid is transferred. Pour the excess solid and the remaining solution down the drain with plenty of water.

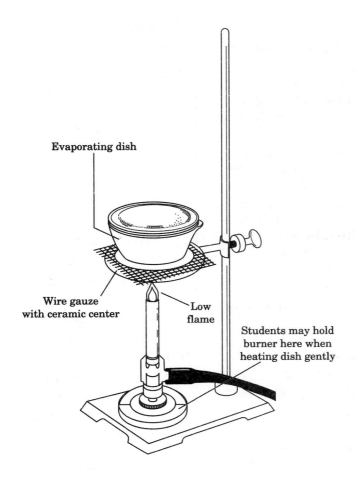

Evaporating dish

Wire gauze
with ceramic center

Low
flame

Students may hold
burner here when
heating dish gently

4. Using tongs, place the evaporating dish on the balance and measure its mass to the nearest 0.01 g. Record the mass in your data table.

5. Using a graduated cylinder, measure approximately 10.0 mL of the salt solution. Record the volume to the nearest 0.1 mL in your data table. Pour this sample of saturated NaCl solution into the evaporating dish.

6. Set up the ring stand assembly and Bunsen burner as shown. The ring and wire gauze should be positioned several centimeters above the burner flame. Place the evaporating dish on the wire gauze, and light the burner. Heat the NaCl solution slowly to avoid spattering. Continue heating until all of the water has evaporated and the remaining crystals are dry.

7. When the remaining crystals appear dry, adjust the burner to produce a very low flame and continue heating for another 10 min. Hold the burner by the base, and gently wave the flame under the evaporating dish, being careful not to heat the material so rapidly that it spatters.

8. Using tongs, place the hot evaporating dish on a piece of wire gauze at the base of the ring stand and allow it to cool. Then determine the mass of the evaporating dish and the solid residue to the nearest 0.01 g. Record the mass in your data table.

9. Measure 5.0 mL of saturated NaCl solution, and pour it into a test tube.

10. Using a spatula, measure about 0.50 g $Na_3C_6H_5O_7 \cdot 2H_2O$ on the balance. Record the exact mass in your data table.

11. Add the $Na_3C_6H_5O_7 \cdot 2H_2O$ to the test tube and stir. Record your observations.

Cleanup and Disposal

12. All of the solutions in this lab may be disposed of by washing them down the drain. Place any excess NaCl or $Na_3C_6H_5O_7 \cdot 2H_2O$ in the disposal container designated by your teacher.

Analysis and Interpretation

1. **Organizing Data**
   Determine the mass of dry NaCl in the evaporating dish by subtracting the mass of the evaporating dish from the mass of the evaporating dish with the NaCl.

2. **Organizing Data**
   Use the periodic table to determine the molar mass of NaCl. Then calculate the number of moles of NaCl in the evaporating dish.

3. **Analyzing Data**
   Calculate the molar concentration of the 10.0 mL of NaCl solution that you evaporated.

4. **Organizing Ideas**
   Write the chemical equation for dissolving NaCl to form the saturated solution.

5. **Organizing Data**
   Use the concentration from item **3** and the equation from item **4** to determine the concentrations of Na^+ and Cl^- ions in the saturated solution.

6. **Applying Ideas**
   Write the equilibrium expression for the equation in item **4.**

7. **Analyzing Data**
   Use your answers to items **5** and **6** to calculate the numerical value of the K_{sp} for NaCl.

8. **Organizing Data**
   How many moles of $Na_3C_6H_5O_7 \cdot 2H_2O$ were added to the solution in step **11?**

Conclusions

9. **Applying Conclusions**
   Use your results from item **3** to determine the mass of NaCl needed to make 4000.0 L of saturated solution.

10. **Analyzing Information**
    Using the equilibrium expression from item **6,** explain your observations from step **11,** when $Na_3C_6H_5O_7 \cdot 2H_2O$ was added to the NaCl solution.

11. **Predicting Outcomes**
    The development team at Juliette foods is also considering adding the following flavoring or preservative additives to the brine: acetic acid, aluminum sulfate, calcium chloride, sodium bicarbonate, and sodium nitrate. Indicate which ones will cause a change in the NaCl equilibrium concentration.

Extensions

1. **Applying Conclusions**
   Suppose your client diluted the saturated NaCl brine to one half of its molarity. Would it be possible then to add 50.0 g of $CaCl_2$ to every liter of the diluted solution without causing precipitation? Show calculations to support your answer.

2. **Applying Ideas**
   Although iron(III) nitrate and iron(III) acetate have very large solubility product constants, farmers do not find these compounds useful as additives for iron-deficient soils. Explain why this is true.

3. **Applying Ideas**
   Assuming that the value for K_{sp} of NaCl stays the same and that all of the $Na_3C_6H_5O_7 \cdot 2H_2O$ dissolves, you can calculate the new equilibrium concentration for Na^+ and Cl^- using the following steps.
   - Calculate the total possible concentration of Na^+ ions.
   - Assume that x mol/L of NaCl precipitates.

- Rewrite the equilibrium expression with $[Na^+]$ equal to the total possible concentration minus x and $[Cl^-]$ equal to the initial concentration minus x.
- Algebraically rearrange the equation until there is a zero on one side of the equation.
- Apply the quadratic formula to the equation from the previous step to determine the value of x, the equilibrium concentration of NaCl. For any equation with the form given below, x has the values shown.

$$ax^2 + bx + c = 0$$

$$x = \frac{-b \pm \sqrt{b^2 - 4ac}}{2a}$$

4. **Research and Communication**
   The hollowing out of limestone caves and the growth of stalactites and stalagmites are environmental examples of solubility equilibria at work. Investigate these processes in terms of equilibrium, and prepare a paper explaining the formation of caves.

5. **Applying Ideas**
   The method for producing sweet pickles is essentially the same as that for making dill pickles, except that the cucumbers are soaked in a saturated sucrose solution instead of a NaCl solution. (Incidentally, sweet pickles were invented when the person doing the pickling mistakenly put sugar into the water instead of salt.) Depending upon the success of their new line of dill pickles, Juliette Brand Foods is planning to expand production to include sweet pickles. Based on your knowledge of how sugar dissolves in water, would you be able to experimentally determine a solubility-product constant for sucrose? Does this affect the precipitation of sugar in the presence of a common ion? Explain your answers.

6. **Predicting Outcomes**
   Because they are just beginning to enter this market, Juliette Brand Foods does not want to have to spend more capital than necessary on equipment. Included in this would be a temperature control for the brine vats, so that the solution remains at a constant temperature at all times. Does the solubility of NaCl vary sufficiently over a range of temperatures to make such a temperature control device necessary? The dependence of solubility on temperature is shown for several solutes in Figure 11-5 on page 407 of the text. Use this diagram to infer how the solubility product constant for NaCl varies with temperature.

7. **Relating Ideas**
   Often it is convenient to express the solubility product constant as pK_{sp}, which is defined similarly to pH.

$$pK_{sp} = -\log K_{sp}$$

   From this definition, what is the value of K_{sp} for a pK_{sp} value of 2.5? 25.7? 132?

Solubility Product Constant—Algae Blooms

April 18, 1995

Ms. Sandra Fernandez
Director of Development
CheMystery Labs, Inc.
52 Fulton Street
Springfield, VA 22150

Dear Ms. Fernandez:

Recently, our region has experienced above-average temperatures and rainfall. These factors have increased runoff into local waterways, causing abnormal algae blooms in lakes and ponds. Our studies show that algae can be controlled with 0.0500 M solutions of copper(II) ions, so we are considering treating affected ponds with copper(II) sulfate or copper(II) chloride.

We would like to apply them to the ponds and lakes in the form of concentrated solutions. In this way, they will mix thoroughly with the lake or pond water much more quickly than if we added the solid compounds.

We are requesting bids for several comparative studies of the solubility properties of these two compounds. We want each contractor to recommend one compound or an optimum combination of both for our use. Please base any calculations of amounts necessary on a 3.90×10^6 L pond and submit experimental data to support your conclusions.

Sincerely,

Kathleen Farros-Hoeppner

Kathleen Farros-Hoeppner
Assistant Director, Research
State Department of Fish and Game

Memorandum

CheMystery Labs, Inc.
52 Fulton Street
Springfield, VA 22150

Date: April 19, 1995
To: Gary Vasileyev
From: Sandra Fernandez

I think the best way to proceed is to <u>measure the</u> <u>concentrations of saturated solutions</u> of each compound. Make a 25 mL saturated solution of each one and <u>calculate K_{sp}</u> so that our results can be compared quickly to those of other firms.

If we land one of these contracts, it will be our <u>first project</u> for a <u>state agency</u>, so we want to make a favorable first impression. Send Ms. Farros-Hoeppner a bid that includes the following.
- detailed <u>one-page plan</u> for the procedure
- examples of all necessary <u>data tables</u>
- list of the <u>materials</u> you need with the <u>total cost</u>

Get started <u>as soon as your plan is approved</u>. When the work is complete, prepare a report in the form of a <u>two-page letter</u> to Kathleen Farros-Hoeppner. The letter must include the following items.
- your <u>recommendation</u> of which compound to kill the algae (be sure to consider costs of each one)
- <u>volumes</u> of each saturated solution needed to achieve a concentration of 0.05 M Cu(II) ions in the pond
- <u>mass</u> of each compound needed to make the solutions
- <u>equilibrium molar concentrations</u> of saturated solutions
- <u>experimentally measured</u> K_{sp} of copper(II) sulfate and copper(II) chloride based on experimental data
- brief description of the <u>procedure</u>
- detailed and organized <u>data table</u> and calculations
- detailed <u>invoice</u> for materials and services

Required Precautions

 Always wear goggles and a lab apron to provide protection for your eyes and clothing.

 When you use a Bunsen burner, tie back loose clothing and hair. Do not heat glassware that is broken, chipped, or cracked. Use tongs whenever handling glassware or other equipment that has been heated.

 Do not touch any chemicals. If you get a chemical on your skin or clothing, wash it off at the sink while calling to your teacher.

 Always wash your hands thoroughly when finished.

Spill Procedures/ Waste Disposal Methods

- In case of spills, follow your teacher's instructions.
- Dispose of the solid wastes in the waste container designated by your teacher.
- Dispose of the liquid wastes in the waste container designated by your teacher.

MATERIALS for State Department of Fish and Game
REQUIRED ITEMS
(You must include all of these in your budget.)

Lab space/fume hood/utilities	15 000/day
Standard disposal fee	2 000/g of product
Balance	5 000
Crucible tongs	2 000
Evaporating dish	1 000

REAGENTS AND ADDITIONAL EQUIPMENT
(Include in your budget only what you'll need.)

$CuCl_2 \cdot 2H_2O$	200/g
$CuSO_4 \cdot 5H_2O$	100/g
150 mL beaker	1 000
400 mL beaker	2 000
250 mL flask	1 000
25 mL graduated cylinder	1 000
100 mL graduated cylinder	1 000
Filter paper	500/piece
Glass stirring rod	1 000
Bunsen burner/related equipment	10 000
Hot plate	8 000
Litmus paper	1 000/piece
Ring stand/ring/wire gauze	2 000
Ruler	500
Spatula	500
Test tube (large)	1 000
Test tube (small)	500
Wash bottle	500

* No refunds on returned chemicals or unused equipment.

FINES

OSHA safety violation	2 000/incident

References

Refer to pages 408–409 in Chapter 11 of the textbook for information about saturated solutions. Page 411 has information about molarity. On pages 462–464 of Chapter 12, you will find background information about equilibrium systems. A discussion of solubility equilibria and the equilibrium constant, K_{sp}, is found on pages 475–478.

Hint: because these compounds can form hydrates, $CuCl_2 \cdot 2H_2O$ and $CuSO_4 \cdot 5H_2O$, they should be heated thoroughly to produce the anhydrous compounds after the solvent is evaporated. $CuSO_4$ should be gray or white (not blue or yellow), and $CuCl_2$ should be light brown.

12-2 Freezing-Point Depression—Testing De-icing Chemicals

Objectives

Measure freezing-point depression for three solutions of unknown solutes.

Calculate the molality of the particles in the solutions from freezing-point data.

Demonstrate how ionic and non-ionic solutes differ in their effect on the freezing point of the solvent.

Determine the best solute for melting ice on roads.

Situation

The Department of Transportation (DOT) in your town is working to minimize hazardous driving conditions during the winter. They are considering adding a soluble substance to the sand that they spread on roads to prevent skidding. The purpose of the soluble material is to lower the freezing point so that the ice will melt. DOT has asked your firm to investigate several potential de-icing agents. They need quantitative freezing-point depression data and your recommendation of which solute is the most effective agent.

Background

Freezing-point depression is one of the colligative properties of solutions discussed in Chapter 12. Colligative properties depend on the number of particles present in a solution. Because ionic solutes dissociate into ions, they have a greater effect on freezing point than molecular solids of the same molal concentration. For example, the freezing point of water is lowered by 1.86°C with the addition of any nonvolatile molecular solute at a concentration of 1 molal. However, a 1 molal sodium chloride solution contains a 2 molal concentration of ions. Thus, the freezing-point depression for NaCl is 3.72°C, double that of a molecular solute. The relationship is given by the following equation.

$$\Delta t_f = K_f \times m \times n$$

Δt_f is the freezing-point depression of the solution, K_f is the molal freezing-point constant (-1.86°C/m for aqueous solutions), m is the molality of the solution, and n is the number of particles formed from the dissociation of each formula unit.

Problem

Other workers have prepared and frozen solutions containing 100.0 g of each de-icing chemical dissolved in 1 kg of water. To make a recommendation to DOT, you must do the following.
• Measure the freezing point of pure water.
• Measure the freezing-point depression for each solution.
• Calculate the molality of particles.

Safety

Always wear goggles and a lab apron to provide protection for your eyes and clothing. If you get a chemical in your eyes, immediately flush it out at the eyewash station while calling to your teacher. Know the locations of the emergency lab shower and eyewash and how to use them.

Do not touch any chemicals. If you get a chemical on your skin or clothing, wash it off at the sink while calling to your teacher. Make sure you carefully read the labels and follow the directions on all containers of chemicals that you use. Do not taste any chemicals or items used in the laboratory. Never return leftovers to their original containers; take only small amounts to avoid wasting supplies.

Always clean up the lab and all equipment after use, and dispose of substances according to proper disposal methods. Wash your hands thoroughly before you leave the lab after all lab work is finished.

Preparation

1. Organizing Data

Create a table in your lab notebook with two columns and five rows. Label the first and second boxes of the first row *Substance* and *Temp. (°C)*. Label the remaining boxes of the first column H_2O, *F-38*, *H-22*, and *J-27*.

Technique

2. Fill a 250 mL beaker with crushed ice. Add distilled water to make a slurry of ice and water.

3. Measure the temperature of the ice-water mixture by holding a thermometer or LEAP System thermistor probe in the mixture. Be sure to keep the instrument off the bottom of the beaker. Wait until the temperature becomes constant, and then read and record the freezing temperature of pure water to the nearest 0.1°C in your data table.

4. Obtain a piece of frozen *F-38* solution that is about 25 mL in volume. Wrap it in toweling, and smash the toweling on the floor two or three times until it has the consistency of crushed ice. Pour the fragments into a test tube.

5. Allow the solid in the test tube to melt sufficiently so that liquid and solid are both present. Stir with a stirring rod to mix the solid and liquid. After stirring, hold a thermometer or LEAP System thermistor probe in the mixture but do not allow it to rest on the bottom of the test tube. When the temperature is constant, read and record the freezing point of the *F-38* solution.

6. Rinse the thermometer and the stirring rod thoroughly. Dry each with a paper towel.

7. Repeat steps **4–6** for the frozen *H-22* and *J-27* solutions. Record the freezing points of these two solutions in your data table.

Materials

- De-icer F-38 (frozen solution)
- De-icer H-22 (frozen solution)
- De-icer J-27 (frozen solution)
- Ice (frozen distilled water)
- 250 mL beaker
- 25 mL graduated cylinder
- Glass stirring rod
- Test tubes (large), 3
- Test-tube rack
- Thermometer, non-mercury

LEAP System option

- LEAP System and thermistor probe

Cleanup and Disposal

8. Each solution should be returned to the disposal container that your teacher set out for it. Clean all apparatus and your lab station. Remember to wash your hands thoroughly before leaving the laboratory.

Analysis and Interpretation

1. Organizing Data
Determine the freezing-point depression, Δt_f, for each solution. The freezing-point depression is the difference between the freezing temperature of the solution and the freezing temperature of the pure solvent, water.

2. Analyzing Data
From the freezing-point depression and the value of the molal freezing-point constant, calculate the "effective molality," which is the molality multiplied by the number of moles of particles per mole of solute ($m \times n$) of each solution, using the equation from the Background section.

3. Analyzing Methods
Explain how you can be sure the temperatures you measured were the melting/freezing points of the solutions tested.

4. Analyzing Methods
Would your results change if you used beaker-sized blocks of the frozen solutions? Why or why not?

Conclusions

5. Evaluating Conclusions
The Department of Transportation would like to use as little solute as possible on the roads. They want the greatest depression of freezing point per gram of solute. Which of the three solutes represents the best option for the DOT?

6. Evaluating Conclusions
The Department of Transportation is also interested in the costs involved. Given the unit costs shown in the table below, which substance is the most economical? (Hint: consider which provides the greatest freezing-point depression per dollar.)

Substance	Cost per kg
F-38	$7.85
H-22	$6.20
J-27	$9.30

7. Evaluating Methods
Why was it important to keep the thermometer off the bottom of the beaker in steps **3** and **5**?

8. Evaluating Methods
Why was it necessary to measure the temperature of the ice and pure water mixture, instead of assuming it to be 0.0°C?

Extensions

1. Applying Conclusions

The solute in F-38 has the formula MX_2, and forms three particles on dissolving in water. The solute in H-22 has the formula MX, and forms two particles on dissolving. The solute in J-27 is a molecular solute which does not dissociate. Calculate the apparent molar mass for each solute. (Hint: first calculate the molal concentration of each solution using the information given here to determine n. Because each solution was made from 100.0 g of solute dissolved in 1.0 kg of solvent, you can then determine approximate molar masses.)

2. Applying Models

Will a $1.0 \, m$ copper(II) sulfate solution have the same freezing-point depression as a $1.0 \, m$ copper(II) chloride solution?

3. Connecting Ideas

Calculate the number of water molecules in enough $1.0 \, m$ aqueous solution of sucrose to contain 2.0 mol of sucrose.

4. Designing Experiments

What possible sources of error can you identify with this procedure? If you can think of ways to eliminate them, ask your teacher to approve your plan, and run the procedure again.

5. Research and Communication

Not everyone believes that using salts as a de-icer on roads is a good idea. The subject is controversial in many northern towns where snowfall is heavy. Find out about the pros and cons. When you have considered all points of view, either hold a class debate, or write a personal position paper and present it as an oral report.

6. Applying Ideas

Ethylene glycol, $C_2H_6O_2$, is a molecular compound used as an antifreeze in automobile engines. What is the molal concentration of ethylene glycol in engine cooling water that will protect an engine when the temperature falls to $-30°C$?

7. Applying Ideas

Antifreeze containing ethylene glycol is also needed in automobile engines in extremely hot climates. Using what you know about other colligative properties, explain why this is so.

8. Relating Ideas

Different solvents have different freezing-point depression constants. For benzene, K_f is $-5.12°C/m$. Pure benzene freezes at $5.50°C$. What would be the freezing temperature of a $0.750 \, m$ solution of sucrose in benzene?

INVESTIGATION 12-2
Problem Solving

Freezing-Point Depression— Making Ice Cream

April 6, 1995

Ms. Martha Li-Hsien
Director of Research
CheMystery Labs, Inc.
52 Fulton Street
Springfield, VA 22150

Dear Ms. Li-Hsien:

In our business, it takes more than a great ice cream recipe to be successful. As you may know, when ice cream is made, it must be kept at a temperature below 0°C. Most ice-cream makers achieve these conditions using a saltwater and ice mixture packed around the container.

With our recent acquisition by D & J Food Processing, many of their products are available to us at a reasonable cost, including chemicals such as calcium acetate, $Ca(C_2H_3O_2)_2 \cdot 2H_2O$, and sodium hydrogen carbonate, $NaHCO_3$.

Please investigate the use of these chemicals and sodium chloride, $NaCl$, in saturated solutions. Determine the solubility of each compound, the molar and molal concentrations, the theoretical freezing-point depression for each compound, and the actual temperature of a mixture of the solution and ice.

Sincerely,

Caitlin Cunningham

Caitlin Cunningham, President
Bubba and Hugo's Ice Cream
A Division of D & J Food Processing

Memorandum

CheMystery Labs, Inc.
52 Fulton Street
Springfield, VA 22150

Date: April 7, 1995
To: Vanessa Fanning
From: Martha Li-Hsien

D & J Food Processing could be a great prospect for future contracts. Our capabilities are a good match for their needs, so be sure to perform a careful and thorough analysis so that we will be assured of repeat business with them.

Mr. Cunningham has been negotiating with our lawyers. Instead of our usual arrangement, they are offering us $15 000 plus expenses for our analysis. Before we can sign the contract, we need the following items from you.

- detailed one-page summary of your procedure
- data charts and tables to organize your information
- itemized expense sheet including total expenditures

As soon as you get approval, begin your work. After your investigation, prepare a report in the form of a two-page letter to Mr. Cunningham at Bubba and Hugo's. Be certain that your report includes the following items.

- theoretically calculated and experimentally measured freezing-point depressions for 50.0 mL of saturated solution for each solute
- percentage difference between the calculated and measured freezing-point depressions
- molal and molar concentrations of the saturated solutions
- values of K_{sp} for each compound
- detailed invoice for $15 000 more than actual total expenditures (be sure to include all expenses so that we will be reimbursed)

Required Precautions

 Always wear goggles and a lab apron to provide protection for your eyes and clothing.

 The ice mixture can cause frostbite if you handle it for too long. Use tongs, towels, or insulated gloves when handling the containers.

 Confine any long hair or loose clothing. Use tongs to handle any glassware or equipment that has been heated.

 Do not touch any chemicals. If you get a chemical on your skin or clothing, wash it off at the sink while calling to your teacher.

 Always wash your hands thoroughly when finished.

Spill Procedures/ Waste Disposal Methods

- In case of spills, follow your teacher's instructions.
- Discard the solutions from this lab by flushing them down the drain.
- Clean the area and all equipment.

References

Refer to pages 408–409 in the textbook for information about saturated solutions and to page 411 for molarity and molality. Refer to pages 459–460 of the textbook for a discussion of freezing-point depression and other colligative properties. On pages 462–464 of Chapter 12, you will find background information about equilibrium systems. A discussion of solubility equilibria and the equilibrium constant, K_{sp}, is found on pages 475–478.

Note that this Investigation requires skills that you used in measuring freezing-point depression and in measuring the solubility product constant for a dissolved substance.

In order to determine molarity and molality of the saturated solutions, it will be necessary to determine the number of moles of each compound, the volume of solution containing that amount, and the mass of the solvent that dissolves that amount.

Base your disposal costs on the mass used for each solute.

MATERIALS for D&J/Bubba and Hugo's Ice Cream
REQUIRED ITEMS
(You must include all of these in your budget.)

Lab space/fume hood/utilities	15 000/day
Standard disposal fee	2 000/g
Balance	5 000
Beaker tongs	1 000
Crucible tongs	2 000

REAGENTS AND ADDITIONAL EQUIPMENT
(Include only what you'll need in your budget.)

$Ca(C_2H_3O_2)_2 \cdot 2H_2O$	500/g
NaCl	500/g
$NaHCO_3$	500/g
Ice	500
250 mL beaker	1 000
400 mL beaker	2 000
250 mL flask	1 000
25 mL graduated cylinder	1 000
100 mL graduated cylinder	1 000
Bunsen burner and related equipment	10 000
Desiccator	3 000
Evaporating dish	1 000
Glass stirring rod	1 000
Hot plate	8 000
Litmus paper	1 000/piece
Paper clips	500/box
Plastic foam cup	1 000
Ring stand/ring/wire gauze	2 000
Stopwatch	2 000
Thermistor probe — LEAP	2 000
Thermometer, nonmercury	2 000
Toothpicks	500/dozen
Wash bottle	500

*No refunds on returned chemicals or unused equipment.

FINES

OSHA safety violation	2 000/incident

Measuring pH—Home Test Kit

April 20, 1995

Mr. Reginald Brown
Director of Materials Testing
CheMystery Labs, Inc.
52 Fulton Street
Springfield, VA 22150

Dear Mr. Brown:

Consumers are becoming more aware of the chemicals around their homes, and they want more information about them. They are asking for home chemical test kits to examine water quality in pools and fish tanks or to check the lead content of a painted wall. Our company would like to tap into this market by developing an inexpensive and easy-to-use pH test kit.

Because any test kit for the home must use very safe materials, we are interested in acid-base indicators that are made from natural products that consumers will recognize. Our kit must cover the pH range from 2 to 12 and be easy to use and interpret. We would like your company to research a wide range of natural indicators and recommend a product for us to test-market within the next six months.

In return, we will pay all reasonable expenses, an additional fee of $25 000, and an option for $\frac{1}{2}$% of the royalties on the first five years' sales of this product.

Sincerely,

Victor Perkes

Victor Perkes
Vice-President
Home Testing Products

Memorandum

Date: April 21, 1995
To: Benito Ramirez
From: Reginald Brown

CheMystery Labs, Inc.
52 Fulton Street
Springfield, VA 22150

If we are successful in this project and the pH test kit sells well, we could be asked to help this start-up company develop other test kits. Unfortunately, our budget is a little tight, and we can afford an up-front investment of only $150 000.

Before you begin, I need the following items to be sent to the company's directors in order to authorize the $150 000 investment.

- detailed one-page plan for your procedure with all necessary data tables
- list of items you will test
- list of the equipment and materials you will need along with the total cost

To get started, pick at least six likely substances to test. I recommend using parts of plants, such as edible fruits and vegetables. No charge will be made for these materials, but you must gather them yourself. For each one, take enough solid starting material to fill a 400 mL beaker to the 250 mL line. Add enough water to fill the beaker. Heat the mixture at about 70°C for about 15 min. Allow it to cool before proceeding. Test the indicator properties of the liquid that remains.

Begin working when your plan is approved. When your team has the answers, prepare a report in the form of a two-page letter to Victor Perkes. The letter must include the following.

- description of the proposed product (solutions, strips, or another method), including your recommendation of the three indicators to be used in the product that together will provide the broadest range and most accurate values
- list of all indicators tested
- detailed and organized data table
- summary paragraph describing your rationale for choosing the indicators
- suggestions for scaling up the manufacturing of the test kit
- suggestions for the wording of instructions for the test kit
- detailed invoice for all expenses and services

Required Precautions

 Always wear goggles and a lab apron to provide protection for your eyes and clothing.

 Do not touch any chemicals. If you get a chemical on your skin or clothing, wash it off at the sink while calling to your teacher.

 Acids and bases are corrosive. If any gets on your skin, wash the area immediately with running water while calling your teacher.

 Confine any long hair or loose clothing. Use tongs to handle any glassware or equipment that has been heated. Allow the heated solution to cool before decanting it or testing its indicator properties.

 Always wash your hands thoroughly when finished.

Spill Procedures/ Waste Disposal Methods

- In case of spills, follow your teacher's instructions.
- Dispose of solid wastes in the container indicated by your teacher. The liquid wastes should be tested for pH. Acidic, neutral, and basic wastes should each go in their own containers.
- Clean the area and all equipment after use.
- Wash your hands when you are finished working in the lab.

References

Pages 508–509 in Chapter 13 of the textbook provide a discussion of indicators and indicator transition intervals, with photos of some indicators as they change color. An explanation of how indicators work is given on page 515. The pH scale and its usefulness in relating the strength of an acid or base solution is discussed on page 501. The dyes in many naturally colored food products can be used as indicators. Red cabbages, berries, and beets contain pH-sensitive chemicals. Do not limit yourself to this list, but try other similar substances.

MATERIALS for Home Testing Products

REQUIRED ITEMS
(You must include all of these in your budget.)

Lab space/fume hood/utilities	15 000/day
Standard disposal fee	2 000/L of solution
Beaker tongs	1 000
Buret clamp, double	1 000
Buret tubes, 2	10 000

REAGENTS AND ADDITIONAL EQUIPMENT
(Include in your budget only what you'll need.)

1.0 M NaOH	100/mL
1.0 M HCl	100/mL
250 mL beaker	1 000
400 mL beaker	2 000
250 mL Erlenmeyer flask	1 000
100 mL graduated cylinder	1 000
Bunsen burner/related equipment	10 000
Glass stirring rod	1 000
Hot plate	8 000
pH paper	2 000/piece
pH meter	3 000
pH probe — LEAP	5 000
Ring stand/ring/wire gauze	2 000
Ruler	500
Stopwatch	5 000
Test tube (large)	1 000
Thermistor probe — LEAP	2 000
Thermometer, nonmercury	2 000
Wash bottle	500

* No refunds on returned chemicals or unused equipment.

FINES

OSHA safety violation	2 000/incident

Measuring pH—Acid Precipitation Testing

Problem Solving

April 24, 1995

Sandra Fernandez
Director of Development
CheMystery Labs, Inc.
52 Fulton Street
Springfield, VA 22150

Dear Ms. Fernandez:

A research team recently discovered that nearby Preston Lake has a pH of 3.0, presumably due to a combination of acid precipitation and point-source pollution. Not surprisingly, many fish were found dead or dying, and the plankton count was far too low. We know that some adult fish begin to die if the pH drops below 5.0, and no fish can reproduce when the pH is below 4.5.

The state legislature recently passed a measure allowing the State Environmental Resources Commission to perform a test of lime neutralization of acidic lakes in the state. We have chosen Preston Lake as the test case.

We are authorized to offer you $100 000 to help us evaluate the use of lime, CaO, to neutralize the lake. Use the sample of lake water we are sending to determine the mass of lime needed for the first treatment of the lake to raise its pH to 6.0 or above. We estimate the volume of Preston Lake to be 2.5×10^9 L.

Sincerely,

Teresa Flores

Teresa Flores
Assistant Director for Water Resources Management
State Environmental Resources Commission

Memorandum

Date: April 25, 1995
To: Erika Strasser
From: Sandra Fernandez

CheMystery Labs, Inc.
52 Fulton Street
Springfield, VA 22150

This project is a promising one. If we do a good job, and if the state legislature decides to fund a full-fledged lake remediation project, we could be called upon to help the State Environmental Re-sources Commission plan treatments for other lakes.

Before you begin work, it is important that you pre-pare the following for the staff at the Environmen-tal Resources Commission to review.

- estimate of the amount of lime you will need to neutralize the 500 mL of lake water, assuming that its pH is 3.0
- detailed one-page plan for your procedure with data tables
- list of necessary equipment and materials with their total cost

After your plan has been approved, begin the lab work. When you have the answers, prepare a report in the form of a two-page letter to Teresa Flores at the State Environmental Resources Commission. The letter must include the following items.

- mass of lime (in kg) that should be used to neu-tralize the 2.5×10^9 L lake
- summary paragraph of how you arrived at your conclusion
- detailed, organized analysis of results with calculations
- organized table of data
- balanced chemical equations for reactions of CaO with the acidic lake water
- discussion of any possible sources of error or difficulties in implementing the plan
- itemized invoice for equipment and services

Required Precautions

 Always wear goggles and a lab apron to provide protection for your eyes and clothing.

 Do not touch any chemicals. If you get a chemical on your skin or clothing, wash it off at the sink while calling to your teacher.

 Acids and bases are corrosive. If any gets on your skin, wash the area immediately with running water while calling your teacher.

 Always wash your hands thoroughly when finished.

Spill Procedures/ Waste Disposal Methods

• In case of spills, follow your teacher's instructions.

• Dispose of solid wastes in the container indicated by your teacher. The liquid wastes should be tested for pH. Acidic, neutral, and basic wastes should each go in their own separate containers.

• Clean the area and all equipment after use.

• Wash your hands when you are finished working in the lab.

References

Acid precipitation often results from the combustion of fossil fuels, a process that can release oxides of nitrogen and sulfur into the atmosphere. These oxides are acid anhydrides. A discussion of the reactions of acid anhydrides with water is found in the textbook on pages 491–493 of Chapter 13. The effects of acid precipitation and neutralization with lime are included in the discussion. Review the pH scale and methods of measuring pH described on pages 499–501. Because of the small amount of sample being tested, it will be most effective if you measure the mass of a beaker with lime before you begin neutralizing. When the pH has reached an appropriate level, record the mass of the beaker with the leftover lime.

MATERIALS for State Environmental Resources Commission

REQUIRED ITEMS
(You must include all of these in your budget.)

Lab space/fume hood/utilities	15 000/day
Standard disposal fee	20 000/L of solution
Balance	5 000

REAGENTS AND ADDITIONAL EQUIPMENT
(Include in your budget only what you'll need.)

CaO, lime	5 000/g
150 mL beaker	1 000
250 mL beaker	1 000
400 mL beaker	2 000
600 mL beaker	3 000
250 mL Erlenmeyer flask	1 000
100 mL graduated cylinder	1 000
Desiccator	3 000
Filter paper	500/piece
Glass stirring rod	1 000
Magnetic stirrer	5 000
Paper clips	500/box
pH meter	3 000
pH paper	2 000/piece
pH probe—LEAP	5 000
Rubber stopper	1 000
Ruler	500
Spatula	500
Tape	500/10 cm

* No refunds on returned chemicals or unused equipment.

FINES

OSHA safety violation	2 000/incident

Acid-Base Titration—Vinegar Tampering Investigation

April 28, 1995

Ms. Marissa Bellinghausen
Director of Investigations
CheMystery Labs, Inc.
52 Fulton Street
Springfield, VA 22150

Dear Ms. Bellinghausen:

Our company manufactures vinegar by the bacterial fermentation of alcohol. This process is capable of producing vinegar containing up to 12% acetic acid, but our practice is to dilute it to a concentration of about 5%, the usual concentration for food products.

Last week, the contract with one of the unions at the plant expired. For a short time, it appeared a strike was imminent, but it was averted at the last minute after an all-night negotiating session. The union has told us they believe that during the uncertainty one of the workers violated plant and union rules by tampering with the dilution equipment. The worker in question left without giving notice after the settlement was announced.

In the meantime, the union and the company agree that it is important to improve our quality assurance to protect consumers from any possible tampering. We would like you to evaluate the sample of vinegar from the production line and send us a report as soon as possible concerning its concentration.

Sincerely,

David Gratton

David Gratton
Director of Quality Control
Indian Summer Vinegar
A Division of D&J Foods

Memorandum

Date: May 1, 1995
To: Michael Bennet
From: Marissa Bellinghausen

CheMystery Labs, Inc.
52 Fulton Street
Springfield, VA 22150

We are building a good relationship with D&J Foods. They're counting on us for quick, accurate results. This time, they are offering us $100 000.

I suggest that you titrate the vinegar. You can choose either a microscale or standard scale titration, but because microscale titrations involve smaller volumes, there is a greater chance for error. Before you start, give me the following items to review.

- detailed one-page plan for the procedure with all necessary data tables
- list of necessary equipment and materials with the total cost

Get started in the lab as soon as your plan has been approved. When you're finished, prepare a report in the form of a two-page letter to David Gratton. The letter must include the following.

- percentage by mass of acetic acid in the sample
- mass of acetic acid in 1.00 L of the sample
- average molarity of the vinegar, based on multiple trials
- pH of the vinegar sample
- K_a of acetic acid
- comparison of and percent difference in your calculation and the accepted value for K_a of acetic acid at 25°C, 1.76×10^{-5}
- summary paragraph describing how you arrived at your answers and discussing any difference between the experimental and accepted values for K_a
- data tables and all calculations
- balanced chemical equation for the titration reaction
- detailed invoice for all materials and services

Required Precautions

 Always wear goggles and a lab apron to provide protection for your eyes and clothing.

 Do not touch any chemicals. If you get a chemical on your skin or clothing, wash it off at the sink while calling to your teacher.

 Acids and bases are corrosive. If any gets on your skin, wash the area immediately with running water while calling your teacher.

 Always wash your hands thoroughly when finished.

Spill Procedures/ Waste Disposal Methods

- In case of spills, follow your teacher's instructions.
- The liquid wastes should be tested for pH. Acidic, neutral, and basic wastes should each go in their own separate containers.
- Clean the area and all equipment after use.
- Wash your hands when you are finished working in the lab.

MATERIALS for Indian Summer Vinegar
REQUIRED ITEMS
(You must include all of these in your budget.)

Lab space/fume hood/utilities	15 000/day
Standard disposal fee	20 000/L of solution

REAGENTS AND ADDITIONAL EQUIPMENT
(Include in your budget only what you'll need.)

1.0 M HCl	500/mL
1.0 M NaOH	500/mL
Phenolphthalein	2 000
150 mL beaker	1 000
250 mL beaker	1 000
125 mL Erlenmeyer flask	1 000
25 mL graduated cylinder	1 000
100 mL graduated cylinder	1 000
Aluminum foil	1 000/cm²
Buret tube	5 000
Filter paper	500/piece
Glass stirring rod	1 000
pH meter	3 000
pH paper	2 000/piece
pH probe—LEAP	5 000
Pipet or medicine dropper	1 000
Ring stand with buret clamp	2 000
Stopwatch	5 000
Test tube (large)	1 000
Wash bottle	500
Watch glass	1 000

* No refunds on returned chemicals or unused equipment.

FINES

OSHA safety violation	2 000/incident

References

Titration, as a way of finding the concentration of an acid solution, is discussed on pages 507–511, and in a How To feature on pages 512–513. Sample Problem 13H on page 514 shows how the molarity of an acid can be calculated from titration data. In Sample Problem 13F on page 504, a method is shown for determining the acid dissociation constant, K_a, of an acid using the pH and molarity of the solution. To determine the mass percentage of acetic acid in vinegar, consider the density of the vinegar to be 1.01 g/mL.

EXPLORATION 14-1 Technique Builder

Catalysts

Objectives

Observe the decomposition of hydrogen peroxide.

Measure the time needed for the production of a given volume of product.

Graph experimental data, and determine the reaction rate.

Infer the effect of a catalyst on the rate of the reaction.

Determine the effect of the concentration of hydrogen peroxide on the reaction rate.

Situation

The winter production of your town's Drama Club will be a stage adaptation of Snow White. In one scene, a large glass flask, filled with what appears to be water, is supposed to begin bubbling and boiling instantly and without a source of heat when the wicked witch drops in a pinch of powdered "wing of bat." The director is looking for a simple, safe chemical reaction that produces a gas and can be initiated by dropping a pellet into a colorless liquid. She has asked you to investigate the decomposition of hydrogen peroxide, H_2O_2, catalyzed by sodium iodide, NaI. She needs data about the reaction rate and how the rate is related to the concentration of both hydrogen peroxide and sodium iodide so that she can adjust it for an effect that will be visible to the audience but safe for everyone on stage.

Background

The measurement of reaction rates and the effects of a catalyst on the rate of reaction are discussed in Chapter 14. The decomposition of hydrogen peroxide to water and oxygen gas is a good illustration.

$$2H_2O_2(aq) \longrightarrow 2H_2O(l) + O_2(g)$$

Without a catalyst, the reaction proceeds slowly, but the rate of reaction is increased dramatically when a catalyst is used. By varying the concentrations of either the reactant or the catalyst, the dependence of the rate on these concentrations can be determined. Rate is measured by determining either the decrease in concentration of a reactant with time or the increase in concentration of a product with time. In this experiment you will use sodium iodide, NaI, as the catalyst and measure the time needed to generate a certain volume of the oxygen product when different concentrations of I^- or H_2O_2 are used.

Problem

To provide the Drama Club with the information they need, you must do the following.
- Carry out the reaction using different concentrations of I^- and H_2O_2.
- Determine the times needed to produce equal volumes of oxygen for different concentrations.
- Graph the data to determine the rate.
- From the graphs, identify a rate law that indicates how the rate is affected by changes in the concentration of H_2O_2 and of the I^- catalyst.

Safety

Always wear goggles and a lab apron to provide protection for your eyes and clothing. If you get a chemical in your eyes, immediately flush it out at the eyewash station while calling to your teacher. Know the locations of the emergency lab shower and eyewash and how to use them.

Do not touch any chemicals. If you get a chemical on your skin or clothing, wash it off at the sink while calling to your teacher. Carefully read the labels and follow the directions on all containers of chemicals that you use. Do not taste any chemicals or items used in the laboratory. Never return leftovers to their original containers; take only small amounts to avoid wasting supplies.

Never put broken glass in a regular waste container. Broken glass should be disposed of separately according to your teacher's instructions.

Always clean up the lab and all equipment after use, and dispose of substances according to proper disposal methods. Wash your hands thoroughly before you leave the lab after all lab work is finished.

Preparation

1. Organizing Data

Prepare a data table with 4 columns and 17 rows. In the first row, label the boxes in the second through fourth columns *Rxn 1 (s), Rxn 2 (s),* and *Rxn 3 (s)*. In the second through fourth rows of the first column, label the boxes *NaI (g), 3% H_2O_2 (mL),* and *H_2O (mL)*. Fill in the remaining 13 rows of this column with even numbers of milliliters of O_2 from 2 to 26. The three remaining columns provide space for recording time in seconds for three reactions. You will also need a space to record the initial volume of the buret for each of the three reactions and several spaces for recording the initial temperature of the water in the pneumatic trough or plastic tub.

2. On separate pieces of weighing paper, measure the mass of three samples of NaI to the nearest 0.01 g. The sample for reaction 1 should be about 0.10 g. The sample for reaction 2 should be about 0.20 g. The sample for reaction 3 should be about 0.10 g. Label each sample with the number of the reaction in which it will be used, and record the masses in your data table.

Technique

3. Using a buret clamp, attach the buret to the ring stand. Attach one piece of rubber tubing to the base of the buret. Thread the rubber tubing through a ring that is attached to the ring stand, and then attach the leveling bulb to the other end of the rubber tubing, as shown in the left portion of the illustration on the next page.

Materials

- 3% H_2O_2 solution, 125 mL
- NaI, 0.50 g
- 125 mL Erlenmeyer flask
- 100 mL graduated cylinder
- 100 mL Mohr buret
- Balance
- Buret clamp
- Leveling bulb
- Medicine droppers, 2
- Pneumatic trough or plastic tub
- Ring stand and ring
- Rubber tubing, 2 pieces
- Rubber stopper, one-hole (for Erlenmeyer flask)
- Rubber stopper, one-hole (for Mohr buret)
- Stopwatch or clock with second hand
- Thermometer, non-mercury
- Weighing paper, 3 pieces

LEAP System Option
- LEAP System
- LEAP thermistor probe

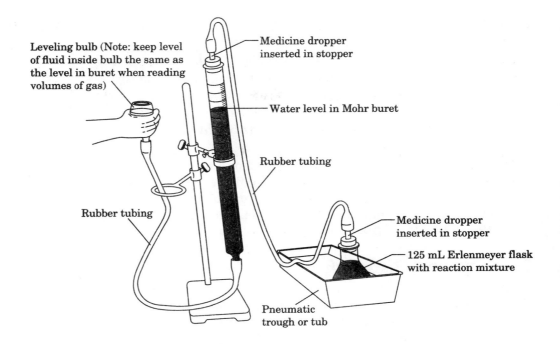

Leveling bulb (Note: keep level of fluid inside bulb the same as the level in buret when reading volumes of gas)

Medicine dropper inserted in stopper

Water level in Mohr buret

Rubber tubing

Rubber tubing

Medicine dropper inserted in stopper

125 mL Erlenmeyer flask with reaction mixture

Pneumatic trough or tub

4. Fill the pneumatic trough or tub with water at room temperature. Add room-temperature water to the buret until the water is about 10 mL from the top when the water in the leveling bulb is at the same level as the water in the buret. To maintain constant pressure throughout the experiment, the water in the leveling bulb should be kept at the same level as the water in the buret. (The bulb may rest in the ring clamp while you complete the assembly of the apparatus.)

5. Your teacher should have already inserted the medicine droppers into the one-hole stoppers. Attach a second piece of rubber tubing, with one of the medicine droppers and stoppers on either end. Insert one stopper into the top of the buret and the other into the top of the flask as shown in the illustration.

6. Check the system for leaks by lowering and raising the leveling bulb. There should be only small changes as the leveling bulb is lowered. If there are large changes, the system has a leak that must be corrected.

7. Record the temperature of the water in the pneumatic trough or plastic tub.

8. Using a graduated cylinder, measure about 50 mL of 3% H_2O_2. Measure the actual volume to the nearest 0.1 mL, and record it in your data table.

9. Record the initial volume of the buret, and decide how you and your partner will divide the tasks in step 10.

10. Remove the stopper assembly from the flask, and place the NaI for reaction 1 into it. Add the hydrogen peroxide, and quickly replace the stopper assembly. One member of your team should swirl the flask throughout the reaction and watch the time. The other member should keep the water in the leveling bulb at the same level as the water in the buret and watch the volume. Begin recording the time and volume after approximately 6 mL of oxygen gas have evolved.

11. Take time and volume readings after the evolution of each 2 mL of gas until you have about 13 readings.

12. Thoroughly rinse the Erlenmeyer flask. Check the temperature of the water in the trough, and, if necessary, adjust it to the temperature recorded in step **7** by adding small amounts of cold or warm water.

13. Repeat steps **8–12** for reaction 2, using 50 mL of 3% H_2O_2 and 0.20 g NaI.

14. Repeat steps **9–11** for reaction 3, but this time add 25 mL of distilled water to the 0.10 g NaI in the flask, and swirl it in the trough until the NaI is completely dissolved and the solution has had time to reach the same temperature as the water bath. Then add 25 mL of 3% H_2O_2, and proceed as before.

Cleanup and Disposal

15. All solutions may be washed down the drain once they have stopped bubbling. Any excess solid NaI should be placed in the disposal container designated by your teacher. Clean up all equipment and your lab station. Remember to wash your hands.

Analysis and Interpretation

1. **Evaluating Data**
   List the five factors that influence the rate of a reaction, and identify those that were studied in this experiment.

2. **Organizing Data**
   Plot your results for reactions 1, 2, and 3 on one graph, with the volume of O_2 in milliliters on the y-axis and the time in seconds on the x-axis. Draw the line of best fit through the points for each reaction.

3. **Interpreting Graphs**
   Calculate the slope for each line. How does it relate to the reaction rate? (Hint: be sure to include the appropriate units.)

4. **Evaluating Methods**
   Why was it important to check the water temperature and keep the bulb's water level the same as the water level in the Mohr buret? (Hint: recall what you know about pressure from Chapter 10 of the textbook.)

Conclusions

5. **Inferring Conclusions**
   Use your graph and your calculated slopes for reactions 1 and 2 to determine how doubling the concentration of the catalyst, NaI, affected the rate of the reaction.

6. **Inferring Conclusions**
   Consider reactions 1 and 3. What effect did the change in concentration of hydrogen peroxide have on the rate of the reaction? (Hint: explain the effect quantitatively.)

7. **Applying Conclusions**
   Write a rate-law expression for this equation using the concentrations of H_2O_2 and I^-. (Hint: use the best value to within 0.1 for the exponents of the concentrations.)

8. Evaluating Data
In this experiment you are studying the rate of a reaction by measuring the increase in volume of oxygen per second rather than the increase in concentration. Explain why volumes of gas, measured under the same conditions of temperature and pressure, can be used instead of concentrations in a rate equation. (Hint: consider what you know about gases from Ch. 10 of the textbook.)

9. Applying Models
Using collision theory, explain why increasing the concentration of a reactant often increases the rate of the reaction.

10. Applying Models
Under what circumstances might increasing the concentration of a reactant have no effect on the rate of a reaction?

11. Applying Conclusions
What concentrations of reactants do you recommend that the Drama Club use for the scene, given that they will be using a 1.0 L round-bottom flask?

Extensions

1. Designing Experiments
What difficulties or sources of error can you identify with this procedure? If you can think of ways to eliminate them, ask your teacher to approve your plan, and run the procedure again.

2. Analyzing Data
The following data were collected in another hydrogen peroxide experiment. What is the average rate of reaction for each of the periods between measurements?

Volume O_2 (mL)	0.0	2.0	3.8	5.9	7.9
Time (s)	0.0	60.0	115.0	180.0	245.0

3. Predicting Outcomes
Predict what the rate might be if the reaction in item **2** continued until 10 mL of O_2 were formed. Explain your prediction.

4. Connecting Ideas
Explain in terms of activation energy for a reaction how a catalyst operates to increase the rate of a reaction.

5. Applying Ideas
Consider the five factors that control the rate of a reaction, and explain why there is a danger of explosion in saw mills and grain elevators, where large amounts of dry, powdered, combustible materials might be present in the air.

6. Research and Communication
Some metabolic functions produce hydrogen peroxide as a byproduct. When present in high concentration, it can cause severe damage to tissue. Find out the mechanisms your body uses to deal with this potentially dangerous substance, and write a short paper summarizing your research.

7. Evaluating Methods
Would you recommend that the Drama Club use this reaction or the reaction of hydrochloric acid with magnesium metal described in Ch.14 of the textbook? Explain your answer. (Hint: remember that the criteria for selecting a reaction are visibility and safety. **Do not try the Mg-HCl reaction.**)

Catalysts — Peroxide Disposal

May 4, 1995

Mr. Reginald Brown
Director of Materials Testing
CheMystery Labs, Inc.
52 Fulton Street
Springfield, VA 22150

Dear Mr. Brown:

In the synthesis of one of our products, hydrogen peroxide is generated as a byproduct. Disposing of the H_2O_2 in the sewer system is not permitted because it can be harmful to wildlife and can cause excess pressure to build up in clogged pipes. We are considering decomposing the hydrogen peroxide into oxygen and water but need a catalyst to speed up the reaction.

We need your firm to investigate three likely catalysts that could be used to speed decomposition of H_2O_2: sodium iodide, manganese dioxide, and yeast. Sodium iodide and manganese dioxide are inorganic catalysts. Yeast is a fungus that produces enzymes which catalyze the decomposition.

Please identify the most effective and economical method for us to use. We are prepared to pay up to $100 000 for satisfactory work.

Sincerely,

Christopher Sasseen

Christopher Sasseen
Division Manager
Industrial Engineering Systems, Inc.

Memorandum

Date: May 5, 1995
To: Theodore Cole
From: Reginald Brown

CheMystery Labs, Inc.
52 Fulton Street
Springfield, VA 22150

We've got to give this contract our best shot because Industrial Engineering Systems is a huge conglomerate and a series of contracts with them could give us the stability we need to further expand our customer base.

Before you make a plan, check the solubility of NaI and MnO_2 in a chemical handbook to be sure they will be satisfactory. Test a small amount of yeast to make sure it is satisfactory. Also be sure to note the costs of all three catalysts. Then, before you begin work in the lab, I will need to review the following items.

- one-page plan for your procedure, with data tables
- detailed list of the equipment and materials you need, with total cost

Begin the lab work as soon as your plan is approved. When you have the answers, prepare a two-page letter to Christopher Sasseen, Division Manager, Industrial Engineering Systems, Inc. Be sure to include the following information.

- recommendation of which catalyst is the most effective and why
- summary paragraph explaining how your procedure solved this problem
- rates of decomposition of H_2O_2 using the three catalysts
- graphs of your data
- detailed, organized data table
- detailed invoice for costs of materials and services

Required Precautions

 Always wear goggles and a lab apron to provide protection for your eyes and clothing.

 Do not touch any chemicals. If you get a chemical on your skin or clothing, wash it off at the sink while calling to your teacher.

 Never put broken glass in a regular waste container. Broken glass should be disposed of separately. Put the glass pieces in the container designated by your teacher.

 Always wash your hands thoroughly when finished.

Spill Procedures/ Waste Disposal Methods

- In case of spills, follow your teacher's instructions.
- MnO_2 and any H_2O_2-MnO_2 mixtures should be placed in a designated and labeled disposal container. **Do not try to wash MnO_2 down the drain.**
- Pure H_2O_2, yeast, and NaI solutions may be poured down the drain.
- Dispose of any leftover solids in the separate disposal containers designated by your teacher.
- Clean the area and all equipment after use.

References

Refer to pages 536–540 for a discussion of reaction rate. The dependence of the rate on the concentration of the reactants is discussed on pages 552–553, and the effects of catalysts are discussed on pages 559–562. Information on measuring the rate of a reaction and graphing rate data to determine the rate law can be found on pages 555–557. The procedure used to measure the rate of decomposition of H_2O_2 will be very similar to one in an Exploration that you and your team recently completed.

MATERIALS for Industrial Engineering Systems, Inc.

REQUIRED ITEMS
(You must include all of these in your budget.)

Lab space/fume hood/utilities	15 000/day
Standard disposal fee	20 000/L of solution
100 mL Mohr buret	5 000
Balance	5 000
Buret clamp	1 000
Leveling bulb	500
Pneumatic trough or plastic tub	1 000
Ring stand/ring	2 000

REAGENTS AND ADDITIONAL EQUIPMENT
(Include in your budget only what you'll need.)

NaI	750/g
MnO_2	150/g
Yeast	100/g
3% H_2O_2 solution	500/mL
Ice	500
250 mL beaker	1 000
100 mL graduated cylinder	1 000
125 mL Erlenmeyer flask	1 000
Beaker tongs	1 000
Filter paper	500/piece
Glass stirring rod	1 000
Rubber stopper with medicine dropper inserted	1 000
Rubber tubing	500/cm
Stopwatch	5 000
Thermometer	2 000
Thermistor probe—LEAP	2 000
Wash bottle	500
Weighing paper	500/piece

* No refunds on returned chemicals or unused equipment.

FINES

OSHA safety violation	2 000/incident

Electroplating for Corrosion Protection

Objectives

Construct an electrolytic cell with an electrolyte, two electrodes, and a battery.

Use the electrolytic cell to plate one metal onto another.

Test the plated metals and a sample of the original metal to determine how well they resist corrosion by an acidic solution.

Relate the results to the activity series and to standard reduction potentials.

Situation

Your company has been contacted by a manufacturer of electrical circuits. The company uses 1.0 M HCl to clean newly manufactured circuits. They've decided to store the acid in large metal tanks instead of reagent bottles. The company has narrowed down the choices to copper, zinc-plated copper, and iron-plated copper. You have been asked to evaluate these choices.

Background

Solutions of acids can oxidize some metals. The single displacement reaction of magnesium and hydrochloric acid is an example.

$$\underset{0}{Mg(s)} + \underset{+1 \ -1}{2HCl(aq)} \longrightarrow \underset{+2 \ -1}{MgCl_2(aq)} + \underset{0}{H_2(g)}$$

Metals that react can be electroplated with a thin layer of a less reactive metal. In an electroplating cell, electrical energy is used to reduce metal ions in solution, causing it to adhere to the surface of an object functioning as an anode. The illustration shown here summarizes the apparatus and the process.

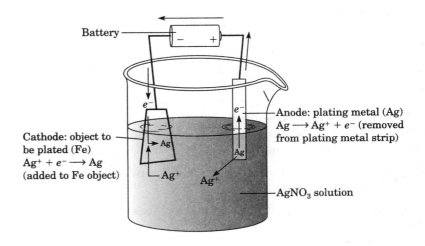

Battery

e^-

Cathode: object to be plated (Fe)
$Ag^+ + e^- \longrightarrow Ag$
(added to Fe object)

Ag

Ag^+

e^-

Ag

Ag^+

Anode: plating metal (Ag)
$Ag \longrightarrow Ag^+ + e^-$ (removed from plating metal strip)

AgNO$_3$ solution

Problem

You must first prepare samples of the plated metals. Then you must test each one in 1.0 M HCl to determine which resists corrosion the best.

Safety

Always wear goggles and a lab apron to provide protection for your eyes and clothing. If you get a chemical in your eyes, immediately flush it out at the eyewash station while calling to your teacher. Know the locations of the emergency lab shower and eyewash station and how to use them.

Do not touch any chemicals. If you get a chemical on your skin or clothing, wash it off at the sink while calling to your teacher. Make sure you carefully read the labels and follow the directions on all containers of chemicals that you use. Never return leftovers to their original containers; take only small amounts to avoid wasting supplies.

Call your teacher in the event of an HCl spill. Acid or base spills should be cleaned up promptly, according to your teacher's instructions.

When you use a Bunsen burner, confine any long hair and loose clothing. Do not heat glassware that is broken, chipped, or cracked. Use tongs or a hot mitt to handle heated glassware or other equipment because hot glassware does not look hot. If your clothing catches on fire, WALK to the emergency lab shower and use it to put out the fire.

Always clean up the lab and all equipment after use, and dispose of substances according to proper disposal methods. Wash your hands thoroughly before you leave the lab after all lab work is finished.

Materials

- 1.0 M HCl, 50 mL
- $FeCl_3$ plating solution, 80 mL
- $ZnSO_4$ plating solution, 80 mL
- Copper wire, 10 cm lengths, 3
- Distilled water
- Iron strip, 1 cm × 8 cm
- Zinc strip, 1 cm × 8 cm
- 150 mL beakers, 3
- 400 mL beaker
- Balance
- Battery, 6 V (lantern type)
- Steel wool
- Stick-on label
- Stopwatch or clock with second hand
- Test-tube rack
- Test tubes, large, 3
- Wax pencil
- Wire with alligator clips, 2 pieces

Optional equipment
- Beaker tongs
- Drying oven

Preparation

1. Organizing Data
Prepare a data table in your lab notebook with four columns and six rows. Label the boxes in the second through fourth columns of the first row *Cu, Fe/Cu,* and *Zn/Cu.* Label the boxes in the second through sixth rows in the first column *Mass of wire (g), Plating time (s), New mass of wire (g), HCl time (s),* and *Mass of wire after HCl (g).* You will also need room to record your observations.

2. Label two of the 150 mL beakers $FeCl_3$ and $ZnSO_4$. Label the 400 mL beaker *Waste.*

3. Make loops on one end of each of the wires. Using a piece of a stick-on label, label the wires *Cu, Fe/Cu,* and *Zn/Cu* just below the loops.

Technique

4. Polish each wire and metal strip with steel wool.

5. Using the laboratory balance, measure the mass of the *Fe/Cu* wire. Record it in your data table.

HRW material copyrighted under notice appearing earlier in this work.

Plating

6. Attach one end of a wire with alligator clips to the loop on the *Fe/Cu* wire. Attach the alligator clip on the other end of the wire to the negative (−) terminal of the battery.

7. Attach one end of the other wire with alligator clips to the iron strip. Clip the other end of this wire to the positive (+) terminal of the battery, as shown in the illustration below. Keep the iron strip away from the copper wire.

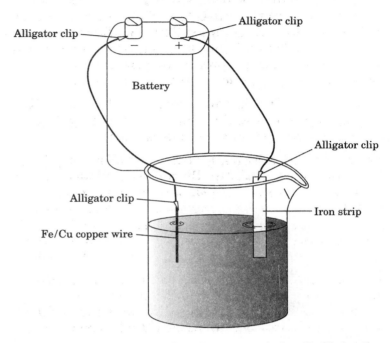

8. Pour 80 mL of the $FeCl_3$ solution into the $FeCl_3$ beaker.

9. Using a stopwatch or clock with second hand to measure the time, immerse both the copper wire and the iron strip in the beaker, being careful to keep the alligator clips out of the solution. Use the loop to hang the wire on the beaker.

10. After about 5 min, remove the copper wire and the iron strip. Record the time elapsed to the nearest 1.0 s.

11. Rinse the *Fe/Cu* wire with distilled water, collecting the rinse water in the *Waste* beaker. Record your observations about the *Fe/Cu* wire in your data table.

12. Hold the unplated *Zn/Cu* wire close to the plated *Fe/Cu* wire. Use a wax pencil to make a mark on the *Zn/Cu* and *Cu* wires at about the same level as the edge of the plating on the *Fe/Cu* wire. Measure and record the masses of these wires.

13. Repeat steps **6–11** with the zinc metal strip, zinc sulfate solution in the $ZnSO_4$ beaker, and the wire labeled *Zn/Cu*. Be certain to plate the wire for exactly the same amount of time. Also be certain that the *Zn/Cu* wire is immersed up to the wax pencil mark.

14. Place the plated wires in the unlabeled 150 mL beaker so that they are not touching each other. Either allow the wires to dry in the beaker overnight, or place the beaker in a drying oven for 10 min. **Remember to use beaker tongs to handle all glassware that has been in the drying oven.**

15. After the wires have cooled, measure and record the masses of the *Fe/Cu* wire and the *Zn/Cu* wire in the data table.

16. In your data table rewrite the original mass of the *Cu* wire in the *New mass of wire (g)* row under the *Cu* column.

Testing Reactivity

17. Fill each of the test tubes about one-third full of 1.0 M HCl. Place the test tubes in a test-tube rack.

18. Place one of the wires into each of the test tubes so that only the plated parts are in the HCl solution. Wait about 5 min.

19. Remove the wires, and rinse them with distilled water, collecting the rinse in the *Waste* beaker. Record the time and your observations about the wires in your data table. Place the wires in the unlabeled 150 mL beaker so that they do not touch. Dry the wires overnight, or place them in a drying oven for 10 min.

20. Remove the beaker of wires from the drying oven and allow it to cool. **Remember to use beaker tongs to handle beakers that have been in the drying oven.**

21. Measure and record the masses of the wires in your data table as *Mass of wire after HCl (g)*.

Cleanup and Disposal

22. Your teacher will provide separate disposal containers for each solution, each metal, and contents of the *Waste* beaker. The HCl from the test tubes can be poured into the same container as the contents of the *Waste* beaker.

Analysis and Interpretation

1. **Applying Models**
   Write the equation for the half-reactions occurring on the metal strip and the copper wire in the *FeCl₃* and *ZnSO₄* beakers. Which is the anode in each one? Which is the cathode in each one?

2. **Analyzing Data**
   How many grams and moles of iron and zinc were plated onto the *Fe/Cu* and *Zn/Cu* wires?

Conclusions

3. **Analyzing Results**
   Which metal or metal combination was the least reactive in HCl? Explain the basis for your conclusion.

4. **Evaluating Conclusions**
   What disadvantages relating to the use of the least reactive metal for the tanks can you think of?

Extensions

1. **Research and Communication**
   Find out what measures are taken to try to prevent metal bridges and buildings from corroding, and prepare a chart to show the different methods, their relative costs, and their general uses.

15-2 *Voltaic Cells*

Technique
Builder

Objectives

Design **and con-
struct various
voltaic cells.**

Measure **the actual
voltages of the
voltaic cells.**

Calculate **the voltages of the voltaic
cells using a table of
standard reduction
potentials.**

Evaluate **cells by
comparing actual
cell voltages and
voltages calculated
from standard re-
duction potentials.**

Write **equations for
cell reactions by
combining half-cell
reactions.**

Situation

NASA is requesting that interested consultants present proposals
for providing electrical engineering services on their "Self-
Propelled Robotic Planetary Explorer." The space agency asks
that your company submit specifications for energy-producing
cells that can be made easily from readily obtainable, inexpensive
materials. Your proposal should provide information about the
construction of the cells, the energy available from the cells, and
the equations for the reactions occurring in the cells.

Background

In Chapter 15, you found that redox reactions involve the transfer
of electrons from a substance being oxidized to a substance being
reduced. If the reactants are in contact, the energy released during the electron transfer will be in the form of heat. However, if
the reactants are in separate half-cells connected by an external
wire, the energy released is electrical energy in the form of electrons traveling through the wire. In electrochemical or voltaic
cells, the oxidation and reduction half-reactions take place in separate half-cells, which often consist of a metal electrode immersed
in a solution of its metal ions. The electrical potential, or voltage,
that develops between the electrodes is a measure of the combined reducing strength of one reactant and oxidizing strength of
the other reactant. This potential difference is measured in units
of volts (V).

Problem

NASA has provided guidelines for constructing and measuring
the potential of a cell using copper and zinc electrodes. To prepare
your proposal you will need to do the following.
- Construct the Cu-Zn voltaic cell, and measure the potential.
- Use standard reduction potentials to determine what other
  cells might be feasible using aluminum, zinc, and copper.
- Prepare two additional cells, and measure their potentials.
- Evaluate the possible choices by comparing the voltages of your
  cells with standard cell potentials.

Safety

Always wear goggles and a lab apron to provide protection for your eyes and clothing. If you get a chemical in your eyes, immediately flush it out at the eyewash station while calling to your teacher. Know the locations of the emergency lab shower and eyewash and how to use them.

Do not touch any chemicals. If you get a chemical on your skin or clothing, wash it off at the sink while calling to your teacher. Make sure you carefully read the labels and follow the directions on all containers of chemicals that you use. Do not taste any chemicals or items used in the laboratory. Never return leftovers to their original containers; take only small amounts to avoid wasting supplies.

Always clean up the lab and all equipment after use, and dispose of substances according to proper disposal methods. Wash your hands thoroughly before you leave the lab after all lab work is finished.

Preparation

1. **Organizing Data**
   Prepare a table for recording your data for three voltaic cells. In each of three columns, you will need a large row to draw a diagram of each cell and regular-sized rows to indicate the molarity of the solutions used and the measured voltage of the cells.

2. Remove any oxide coating from strips of aluminum, copper, and zinc by rubbing them with emery cloth. Keep the metal strips dry until you are ready to use them.

3. Label three 150 mL beakers $Al_2(SO_4)_3$, $CuSO_4$, and $ZnSO_4$. If you are using porous cups, also label a 250 mL beaker $ZnSO_4$ *half-cell*.

Technique

4. If you are using a porous cup, pour 75 mL of 0.5 M $ZnSO_4$ into the $ZnSO_4$ *half-cell* beaker and 75 mL of 0.5 M $CuSO_4$ into one of the porous cups. If you are using a salt bridge, pour 75 mL of 0.5 M $ZnSO_4$ into the $ZnSO_4$ beaker and 75 mL of 0.5 M $CuSO_4$ into the $CuSO_4$ beaker.

5. If you are using a porous cup, place the cup containing $CuSO_4$ solution into the beaker containing the $ZnSO_4$ solution. If you are using a salt bridge, place one end into the solution in the $CuSO_4$ beaker and the other end into the solution in the $ZnSO_4$ beaker.

6. Place a zinc strip into the zinc solution and a copper strip into the copper solution.

Materials

- 0.5 M $Al_2(SO_4)_3$, 75 mL
- 0.5 M $CuSO_4$, 75 mL
- 0.5 M $ZnSO_4$, 75 mL
- Aluminum strip, 1 cm × 8 cm
- Copper strip, 1 cm × 8 cm
- Zinc strip, 1 cm × 8 cm
- Distilled water
- 100 mL graduated cylinder
- Balance
- Emery cloth

Porous Cup Option
- 150 mL beakers, 3
- 250 mL beakers, 3
- Porous cup

Salt Bridge Option
- 150 mL beakers, 3
- Salt bridge

Voltmeter Option
- Voltmeter
- Wires with alligator clips, 2

LEAP System Option
- LEAP System
- LEAP voltage probe

7. Follow the instructions below for the type of voltage-measuring device you will be using.

   a. **Voltmeter**
      Using the alligator clips, connect one wire to the end of the zinc strip and the second wire to the copper strip. Take the free end of the wire attached to the zinc strip, and connect it to one terminal on the voltmeter. Take the free end of the wire attached to the copper strip, and connect it to the other terminal on the voltmeter. The needle on the voltmeter should move to the right. If your voltmeter's needle points to the left, disconnect the wires and reconnect the zinc strip and copper strip to different terminals of the voltmeter. Immediately record the voltage reading in your data table, and disconnect the circuit.

   b. **LEAP System with voltage probe**
      If you are using the LEAP System voltage probe, connect one of the alligator clips to the zinc strip and the other to the copper strip. If the LEAP System voltage probe's wires do not reach far enough, consider using additional wires with alligator clips as described for the voltmeter option. The value on the screen should be positive. If your reading is negative, disconnect the wires, and switch the alligator clips on the LEAP voltage probe to the other metal strips. Immediately record the voltage reading in your data table, and disconnect the circuit.

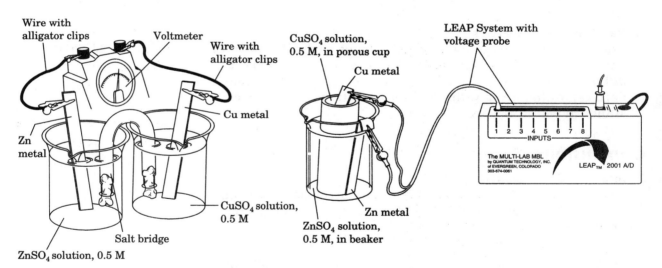

8. Fill in the rest of the data table, including the sketch of the equipment as it appeared when connected.

9. If you are using porous cups, pour each of the solutions into the appropriately labeled 150 mL beakers.

10. Rinse the copper strip with a *very small* amount of distilled water, collecting the rinse in the $CuSO_4$ beaker. Rinse the zinc strip with a *very small* amount of distilled water, collecting the rinse in the $ZnSO_4$ beaker. If a porous cup was used, rinse the outside, collecting the rinse in the $ZnSO_4$ beaker. Then, rinse the inside of the porous cup, collecting the rinse in the $CuSO_4$ beaker. If a salt bridge was used, rinse each end into the corresponding beaker.

11. Use a table of standard reduction potentials, such as **Table 15-1** on page 585 of the textbook, to calculate the standard voltages for the other cells you can build using copper, zinc, or aluminum. Build these cells and measure their potentials according to steps **4–10.**

Cleanup and Disposal

12. Clean all apparatus and your lab station. Remember to wash your hands. Place the pieces of metal in the disposal containers designated by your teacher. Each solution should be poured in its own separate disposal container. Do not mix the contents of the beakers.

Analysis and Interpretation

1. **Organizing Ideas**
   For each cell that you constructed, write the equations for the two half-cell reactions. From a table of standard reduction potentials, obtain the standard half-cell potentials for the half-reactions, and write these E^0 values after the equations.

2. **Organizing Ideas**
   Combine the two half-reactions for each cell you tested to obtain the equations for the total cell reactions.

3. **Organizing Ideas**
   Combine the E^0 values for each cell to find the standard potential for each cell.

4. **Resolving Discrepancies**
   Compare the actual cell voltages you measured with the standard cell voltages in item **3.** Explain why you would expect a difference. (Hint: consider how the construction of the standard cell differs from yours and the conditions under which standard potentials are measured.)

Conclusions

5. **Inferring Conclusions**
   On the basis of the actual voltages that you measured, which cell would you recommend to NASA as the best energy producer?

6. **Applying Ideas**
   On the basis of your data, which metal is the strongest reducing agent? Which metal ion is the strongest oxidizing agent?

7. **Analyzing Methods**
   Why was it important that you removed any oxide coating from the metal electrodes before they were used?

8. **Applying Ideas**
   For each of the diagrams in your data table, indicate the direction of electron flow.

Extensions

1. **Predicting Outcomes**
   Describe how and why the reactions would stop if the cells had been left connected.

2. Evaluating Conclusions

As with any space flight, considerations of mass are critical for the missions in which NASA would use the "Self-Propelled Robotic Planetary Explorer." Given your answer to Extension **1,** which of the cells you tested would continue providing some electrical energy for the longest period, assuming that equal masses of the metals are used in all of the cells?

3. Designing Experiments

Design a method that could use several of the electrochemical cells you constructed to generate more voltage than any individual cell provided. (Hint: consider what would happen if you linked an Al-Zn cell and a Zn-Cu cell. If your teacher approves your plan, test your idea.

4. Applying Ideas

The electrochemical cell shown below is at a temperature of 25°C. Identify the metal that will be oxidized if the current is allowed to flow.

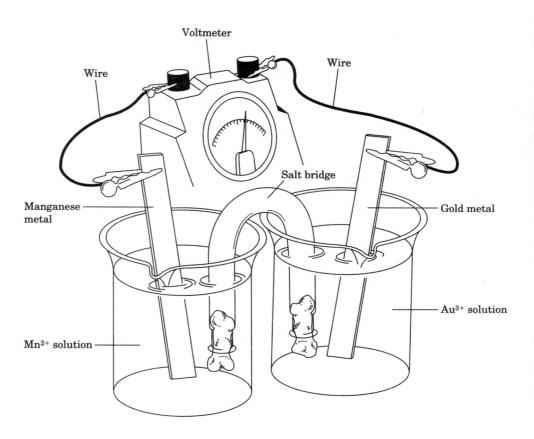

Voltmeter

Wire

Wire

Salt bridge

Manganese metal

Gold metal

Mn^{2+} solution

Au^{3+} solution

5. Research and Communication

Electric automobiles powered by rechargeable batteries are now in limited use in some areas. Find out about the available types of batteries, the range of the vehicle, and the advantages and disadvantages. Prepare a report to your class about your findings, including your predictions about the future of this form of transportation.

INVESTIGATION

15-2

Problem Solving

Voltaic Cells—Designing Batteries

May 9, 1995

Ms. Sandra Fernandez
Director of Development
CheMystery Labs, Inc.
52 Fulton Street
Springfield, VA 22150

Dear Ms. Fernandez:

Rescue Adjuncts, Inc. makes a variety of survival and first-aid gear for outdoor enthusiasts. We want to develop a powerful signal light that can be used by people venturing into areas where ordinary modes of communication are not available. Life or death outcomes depend on whether a distress signal reaches observers who can send help.

We are requesting your help in developing voltaic cells capable of powering a signal light. Please research which of the following metals and their solutions will provide at least 5.0 V: aluminum, copper, magnesium, tin, and zinc.

Submit a proposal that includes the best prospects for the cells. Include a bid that estimates the cost of whatever research you need to do in order to recommend the voltaic cell that we hope will become indispensable to every adventurer.

Sincerely,

Garlon Prewitt

Garlon Prewitt
Chief Engineer
Rescue Adjuncts, Inc.

Memorandum

Date: May 10, 1995
To: Lucinda Crevois
From: Sandra Fernandez

CheMystery Labs, Inc.
52 Fulton Street
Springfield, VA 22150

This job is by no means certain. My contacts elsewhere tell me that this offer went out to at least three of our competitors. If we want to get this business, we must be certain to have the best (and most cost-effective) proposal. As discussed in the letter from Garlon Prewitt, be sure your proposal includes the following.

- detailed one-page plan for your procedure
- combinations that are most likely to yield the highest voltages (maybe several cells linked together)

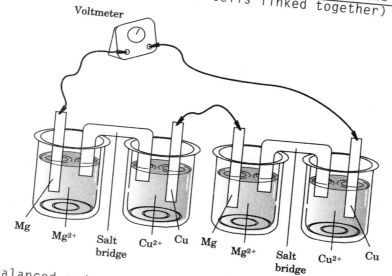

Voltmeter

Mg   Mg²⁺   Salt bridge   Cu²⁺   Cu   Mg   Mg²⁺   Salt bridge   Cu²⁺   Cu

- balanced redox equations and cell diagrams for the voltaic cells you will construct
- standard cell potentials for your proposed cells
- detailed data tables
- list of all necessary materials and equipment
- detailed bid with total costs (Because this is a bid, you need to factor in equipment, materials, labor, overhead, and profit.)

Begin working as soon as your bid is accepted. Make sure you keep your actual costs under the amount of the bid. When your work is finished, prepare a two-page report to Garlon Prewitt. Be sure to include the following items.

- your final recommendation for the cell or combination of cells that will provide 5.0 V
- sketch or diagram of the cell or combination
- paragraph summarizing your procedure
- detailed and organized data and analysis sections
- calculations showing what percentage difference there was between your measured cell potentials and the standard potentials calculated for the proposal
- invoice for the bid amount

Required Precautions

Always wear goggles and a lab apron to provide protection for your eyes and clothing. If you chose to work with the $SnCl_2$ solution, you must wear gloves because the solution irritates the skin.

Do not touch any chemicals. If you get a chemical on your skin or clothing, wash it off at the sink while calling to your teacher.

Never put broken glass in a regular waste container. Broken glass should be disposed of separately. Put the glass pieces in the container designated by your teacher.

Always wash your hands thoroughly when finished.

Spill Procedures/ Waste Disposal Methods

- In case of spills, follow your teacher's instructions.
- After testing a cell, rinse the metal strips (and the porous cup, if one was used) with a small amount of distilled water, collecting the rinse in the beaker containing the solution of that metal's ions.
- **Do not mix the solutions or pour them down the drain.** All solid and liquid wastes must be disposed of in their own separate containers, as designated by your teacher.
- Clean the area and all equipment after use.

References

Spontaneous redox reactions and their application in electrochemical cells are discussed on pages 576–579. Refer to pages 582–584 for information about half-reactions and electrode potentials. A table of standard reduction potentials is found on page 585. See page 586 for examples of how standard reduction potentials are combined to give the total cell potential. The overall voltage of a cell may be enhanced by linking voltaic cells into a series circuit with the cathode of one cell attached with a wire to the anode of another cell. The procedure and the necessary calculations are similar to one your team recently completed using Al, Cu, and Zn. Because the metal strips can be reused, do not include them in your disposal costs.

MATERIALS for Rescue Adjuncts, Inc.

REQUIRED ITEMS
(You must include all of these in your budget.)

Lab space/fume hood/utilities	15 000/day
Standard liquid disposal fee	20 000/L

REAGENTS AND ADDITIONAL EQUIPMENT
(Include in your budget only what you'll need.)

0.5 M $Al_2(SO_4)_3$	100/mL
0.5 M $CuSO_4$	100/mL
0.5 M $MgSO_4$	100/mL
0.5 M $SnCl_2$	100/mL
0.5 M $ZnSO_4$	100/mL
Metal strips: Al, Cu, Mg, Sn, Zn	1 000/each
150 mL beaker	1 000
250 mL beaker	1 000
100 mL graduated cylinder	1 000
Emery cloth	1 000
Filter paper	500/piece
Glass stirring rod	1 000
Porous cup	2 000
Ring stand with buret clamp	2 000
Salt bridge	2 000
Tongs or forceps	500
Voltage probe —LEAP	3 000
Voltmeter	3 000
Wire with alligator clips	500

* No refunds on returned chemicals or unused equipment.

FINES

OSHA safety violation	2 000/incident